New Mexico Mathematics Contest Problem Book

NEW MEXICO MATHEMATICS CONTEST PROBLEM BOOK

LIONG-SHIN HAHN

Foreword by Reuben Hersh

University of New Mexico Press

Albuquerque

To my grandchildren

Bieng-Chi, Ih-Shiu, Bieng-Hui, Ih-An, Bieng-Hi

Printed in the United States of America

Library of Congress Cataloging-in-Publication Data

Hahn, Liong-shin, 1932–
New Mexico mathematics contest problem book / Liong-shin Hahn ; foreword by Reuben Hersh.— 1st ed.
p. cm.
Includes bibliographical references.
ISBN 0-8263-3534-9 (pbk. : alk. paper)
1. Mathematics—Problems, exercises, etc.
2. Mathematics—Competitions—New Mexico. I. Title.
QA43.H246 2005
510'.76—dc22
2005012828

Design: Mina Yamashita
Math text set in LaTeX with illustrations by the author

CONTENTS

Foreword

By Reuben Hersh

Every November, some thousand middle- and high-school students in dozens of schools all over New Mexico gather to spend up to three hours playing and struggling with a set of elementary but tricky math problems. It's Round One of the New Mexico High School Math Contest. Then, the following February, the highest-scoring two hundred or so are invited to the campus of the University of New Mexico in Albuquerque. They hear a talk by a famous mathematician (Paul Erdös, Peter Lax, John Conway, Serge Lang, Bill Thurston, and Ron Graham have done it), have lunch, and then dig in to Round Two, a more rigorous and demanding set of mathematical mind-twisters.

In April, at a banquet in Albuquerque, prizes are awarded by grades from seventh to twelfth. Top prizes are books, money and scholarships. Everybody who does the second round gets a free T-shirt, such as the one I am wearing as I write this.

For many students, this contest has led to steps, decisions, and, ultimately, successful careers in mathematics and science. Some parents sing the praises of an event, uncommon in parts of our state, that publicizes and glorifies the work of the mind (a kind of activity not always and everywhere rewarded among high-school kids in the present day U.S.).

All this takes time and money. At the first round, local teachers volunteer as coaches and proctors. At the second round and for the grading, faculty and graduate students at the University of New Mexico volunteer. The costs of the banquet and the prizes are graciously met by the PNM Foundation. The University of New Mexico contributes the use of its facilities and perhaps part-time release from teaching for a faculty member responsible for organizing all this.

The key element, of course, is the questions themselves. They must be attractive, intriguing, and suggestive to teenage contestants. They must require an insight, a spark, beyond routine application of a standard recipe. They must be solvable with no knowledge beyond that to be expected of a high-school student. And they should be original, not retreads of familiar chestnuts in old problem books. Such are the creations of Liong-shin Hahn—"L.S.," as he allows himself to be called by those who are shy of pronouncing his full name. Rather than cite examples, I invite you to open this book and read from any page.

The New Mexico Mathematics Contest is part of a long tradition with roots in Hungary; the story brings together mathematicians from three continents. The recently retired director of the contest, Professor Liong-shin Hahn, whose remarkable collection of problems you are now holding, was born in 1932 in Taiwan into a family of physicians. He calls himself a "black sheep" for having strayed into mathematics. He got his Ph.D. at Stanford in 1966 with Karel deLeeuw, was an instructor at Johns Hopkins, and came to New Mexico in 1968.

In 1998 Professor Hahn received the "Citation for Public Service" from the American Mathematical Society at their national meeting in Baltimore. The citation says, "Professor Hahn is being honored for carrying forward and developing the New Mexico High School Mathematics Contest and for exposition and popularization of mathematics attractive and suitable for potential candidates for the contest and others with similar intellectual interest." The reference to exposition and publication recognizes Hahn's two excellent books, *Classical Complex Analysis* (with Bernard Epstein) and *Complex Numbers and Geometry*. The latter is an original unique work, easily accessible to high-school students and teachers.

The history of the contest can be traced back to 1894, when a "Pupils' Mathematical Competition" was established by the Mathematical and Physical Society of Hungary. It became one of the decisive influences that made tiny Hungary one of the major contributors to twentieth-century mathematics. For decades the contest was known as the Eötvös Competition, in honor of the Society's founder and president, the physicist and Minister of Education, Baron Loránd

Eötvös. The contest was established at the motion of Gyula König, a powerful personality who dominated Hungarian mathematical life for decades.

Theodore von Kármán, one of the preeminent founders of modern aeronautics, reminisces in his autobiography, *Wind and Beyond*, about high school in Budapest. "Each year the high schools awarded a national prize for excellence in mathematics. It was known as Eötvös Prize. Selected students were kept in a closed room and given difficult mathematics problems, which demanded creative and even daring thinking. The teacher of the pupil who won the prize would gain great distinction, so the competition was keen and teachers worked hard to prepare their best students. I tried out for this prize against students of great attainment, and to my delight I managed to win. Now, I note that more than half of all the famous expatriate Hungarian scientists and almost all the well-known ones in the United States have won this prize."

Besides von Kármán himself, Edward Teller is the prize winner whose name would be most familiar in the U.S. Mathematicians will recognize the names of Lipot Fejér (one of the great contributors to Fourier analysis), Denes König (the son of Gyula, and father of graph theory), Marcel Riesz (a major contributor to partial differential equations, who settled in Sweden and became the mentor of Lars Garding and Lars Hörmander), Alfred Haar (Haar measure, among other things), Tibor Rádo (theory of surface area), and Gabor Szegö (coauthor with George Pólya of one of the great classics in complex analysis).

Von Kármán wrote further, "I think that this kind of contest is vital to our educational system, and I would like to see more such contests encouraged here in the United States and other countries." His book was published in 1967, when von Kármán was a professor at Caltech, in Pasadena, California. His wish was already being carried out on the peninsula south of San Francisco.

Von Kármán was one of many outstanding Jewish mathematicians who fled Europe for America in the 1930s and '40s. Another was Gabor Szegö, winner of the Eötvös Competition in 1912. Szegö became chairman of the Mathematics Department at Stanford University, and in 1946 he initiated the Stanford University Competitive Examination in Mathematics. He had been joined at Stanford by his longtime friend George Pólya, coauthor of the famous *Aufgaben und Lehrsatze der Analysis*. This unique book teaches beginning and advanced methods in complex analysis by a carefully arranged sequence of problems (with hints). Pólya and Szegö wrote, "This book aims to stimulate the reader to independent work. . . . The imparting of factual knowledge is for us a secondary consideration. Above all we aim to promote in the reader a correct attitude, a certain discipline of thought."

This was also one of the deep aims of Pólya's famous work, both in teaching problem solving and teaching the teaching of problem solving. In the introduction to the *Stanford Mathematics Problem Book* (coauthored by Jeremy Kilpatrick), Pólya wrote that the Stanford examination "was established in the belief that an early manifestation of mathematical ability is a definite indication of exceptional intelligence and suitability for intellectual leadership in any field of endeavor." In the introduction to his teacher-training text, *Mathematical Discovery*, he wrote, "Solving problems means finding a way out of a difficulty, a way around an obstacle, attaining an aim which was not immediately attainable. Solving problems is the specific achievement of intelligence, and intelligence is the specific gift of mankind. Solving problems can be regarded as the most characteristically human activity."

The late Professor Abe Hillman (he died in January 2004) brought the Hungarian-style high-school math contest to New Mexico. Hillman was born in Brooklyn, New York, in 1918. He contracted polio in childhood and throughout life had to be assisted by crutches or a wheelchair. At Brooklyn College in 1938, he was one of the top five contestants in the William Lowell Putnam Mathematical Competition for undergraduate college students. (Also among the top five that year was the future famous physicist Richard Feynman.) In 1939 Hillman was a member of the Brooklyn College team that won the Putnam. He earned his Ph.D. at Princeton, under the famous topologist Solomon Lefschetz.

From 1956 to 1965 Hillman was a professor at Santa Clara University in California, where he had opportunities for direct personal contact with Pólya at nearby Stanford. At Santa Clara, Hillman established a high school problem contest that will soon celebrate its fiftieth anniversary. From 1973 to 1977 Hillman was national director of the Putnam Competition. In 1972 the Mathematical Association of America instituted a national high school exam. In 1974 United States high-school students for the first time participated in the International Mathematical Olympiad. They amazed everyone by finishing second, only a few points behind the Soviet Union. When Hillman's death was announced five months ago, I received a message from a senior scientist at Los Alamos National Laboratory. He wrote that his participation in the Santa Clara contest was the inspiration for his successful lifetime career in science.

Hillman retired in 1990. Hahn ran the UNM Contest from 1990 to 1999. Hahn retired in 1999, but is still very much part of New Mexico mathematics. Not only does he continue to help in making up problems, but every February he returns from Irvine, California, to help run the second round of the contest. With this book, he will become a permanent fixture in New Mexico mathematics.

Many former contestants in the New Mexico contest are now leaders in different areas of science and scholarship, including, of course, mathematics and different parts of the computer industry. I would like to mention Bill Duke, also a former student of mine and a recognized leader in number theory, now at UCLA; Brian Conrey, one of the leading researchers on Riemann's hypothesis on the zeroes of the zeta function and executive director of the American Institute of Mathematics in Morgan Hill, California; and astronaut Harrison Schmidt.

In addition to his problems for the New Mexico contest, Hahn has included in his book four of his contests on calculus (lower-division undergraduate college-level) and his annual puzzles on the calendar year all the way up to 2016. These are at the end of the book, and they give you a chance to see a master problem creator at work. Given the number 1997, 1998, or 1999, make up an interesting math puzzle based on that number! Can you do it? How do you start? He has been doing it every year since 1985! If you look at these problems, you see that they involve much more than doing some multiplying or adding that gives you the desired number. There is always surprise, elegance, and ingenuity—in a word, wit! Yes, there is such a thing as witty problem-creating, and Hahn is a great example of it.

Hahn's book will be treasured by problem addicts everywhere. Here in New Mexico it will be the indispensable spur and goad for kids who love math, and, of course, also for their teachers.

References

Hahn, Liang-shin. 1994. *Complex Numbers and Geometry*. Mathematical Association of America.

———, and Bernard Epstein. 1996. *Classical Complex Analysis*. Sudbury, MA: Jones and Bartlett.

Hersh, Reuben and Vera John-Steiner. 1993. "A Visit to Hungarian Mathematics." *Mathematical Intelligencer* 15, no. 2: 13–26.

Olsen, Steve. 2004. *Count Down*. Boston: Houghton Mifflin.

Pólya, George. 1957. *How to Solve It*. Princeton, NJ: Princeton University Press.

———. 1980. *Mathematical Discovery*. New York: John Wiley.

———. 1974. *The Stanford Mathematics Problem Book*. New York: Columbia University Teachers College.

———, and Gabor Szegö. 1925. *Aufgaben und Lehrsatze aus der Analysis*. Berlin: Springer-Verlag

Rádo, Tibor. 1932. "On Mathematical Life in Hungary." *American Mathematical Monthly* 37: 85–90.

Rapaport, Eva. 1963. *Hungarian Problem Book I and II*. New York: Random House.

von Kármán, Theodore and Lee Edson. 1967. *Wind and Beyond*. Boston: Little, Brown.

Preface

For more than thirty years, beginning with the 1966–67 academic year, the Department of Mathematics and Statistics at the University of New Mexico has sponsored the New Mexico Mathematics Contest for secondary school students. Beginning with the 1990–91 academic year (except in the 1996–97 academic year when the contest was suspended), I was solely responsible for composing the contest problems until my retirement in the summer of 1999. Before that, I had contributed problems, off and on, for more than ten years. I believe that a mathematics contest offers a wonderful opportunity to inspire and broaden the horizons of our youngsters (as well as their teachers) and not just to choose winners. Therefore, I would select problems that are not only interesting and challenging, but also of educational value. I want the contestants to learn some mathematics after three hours of intensive thinking and to demonstrate their insight and creativity rather than their stamina. (I do not think problems that require excessive trial and error are suitable for a mathematics contest, especially in this computer age.)

The New Mexico Mathematics Contest has generated growing enthusiasm among students, parents, and teachers throughout New Mexico. Having taken the advice of many friends to publish a selection of my problems, I have chosen my favorite problems (some of them are in my file but were never used) and grouped related problems together for convenience in studying.

The New Mexico Mathematics Contest consists of two rounds. The first round in November is open to all secondary school students in the state. The contest committee selects approximately the top 15 percent of students from each grade and invites them for the final round in February. This format enables me to pair the problems in the two rounds. For example, consider the following: Through a point in a triangle, draw lines parallel to the three sides. That divides the triangle into three small triangles and three parallelograms. In the first round, I would give the areas of the three small triangles and ask the contestants to find the area of the original triangle. Then in the second round, I would give the areas of the three parallelograms and ask the contestants to find the area of the original triangle. In this way I try to encourage students to explore the problems in the first round for possible extensions and variations as they prepare for the final round. In the process, I hope they learn not only mathematics, but also an approach to mathematics.

I have written this book to reflect my teaching philosophy: Students learn best when they discover the result or the method on their own. I like to teach the basic ideas and concepts, then use questions and exercises to guide students to explore and discover for themselves. As a corollary, I am allergic to fat textbooks trying to teach everything. Another corollary is that much of the value in this book is in working through the problems yourself. More specifically, it is very important that you do not read the solution until you have solved the problem yourself. Or, at least you have put in an honest effort to solve it. In solving a problem, put the problem in perspective. Ask yourself:

Do I understand the problem thoroughly?
What is the assumption?
What is the conclusion?
Did I use all the conditions fully?
Can I solve some particular cases first?
Is the assumption sufficient?
What if part of the condition is modified this way or that way?
Is the conclusion reasonable?
What do I need to establish the final conclusion?
What are the consequences if the conclusion is true?
What if the conclusion is wrong?

After, say, twenty-four hours of honest effort and you still cannot solve the problem, then you may peek at the solution. But only the first few lines. As soon as you read an idea or an approach that did not occur to you, then you should close the book immediately, and try to complete the solution yourself. If, after another few hours of grappling, you still cannot solve the problem, then you may read a few more lines of the solution.

After you have succeeded in solving the problem, you should look back and compare your solution with the one in the book. What are the common features? What are the differences? Again ask yourself—

Can I get a stronger result?
Other conclusions?
What if the condition is weakened?
Is the converse true?
That is, can I trade the conclusion with (a part of) the assumption?

If you take this kind of approach to every problem, then your problem-solving ability will surely improve.

I am of the opinion that mathematics is one entity, and that the division of mathematics into algebra, geometry, analysis, etc., is merely for ease and convenience in studying and teaching. Consequently, I am fond of "mixed" problems. As a result, even though I have divided the problems into two categories, Number Theory and Algebra (chapter 1.1) and Geometry and Combinatorics (chapter 1.2), for the convenience of the readers, many problems are difficult to classify. In general, trigonometry problems that involve geometry are in chapter 1.2, and those that do not are in chapter 1.1. Problems that can be solved by similar methods are grouped together. Therefore, if a reader encounters difficulty in solving a problem, he or she is advised to try neighboring problems and then come back to the original problem.

In composing contest problems for over a decade, I have solicited comments before and after each contest from my friends. It is my pleasure to express appreciation to my colleagues, Professors Mutiara Buys, Jeff Davis, Ralph Demarr, Bernard Epstein, Cathy Gosler, Reuben Hersh, Frank Kelly, Dick Metzler, Maria Cristina Pereyra, Tim Berkopec, Takashi Hosoda; and a group of teachers at Lakeside School, Seattle, Washington, especially, Doug Anderson, Dean Ballard, Irene Barinoff, Jabe Blumenthal, Deborah Fogerty, Liz Gallagher, John Jamison, Tom Rona, Paul Stocklin, Fred Wright; also Etsuo Ogiwara of Sapporo, Japan; Nobuyoshi Shimotakahara of Miyazaki, Japan; Mary Newsom of San Antonio, Texas; and my good friends, Professor Tong-Shieng Rhai of National Taiwan University, Taipei, Taiwan, and Professors Kung-Wei Yang and Po-Fang Hsieh, both of Western Michigan University, Kalamazoo. My special thanks are due to Bernard Epstein, who read the entire manuscript meticulously and made valuable suggestions. I can never thank him enough for his help and friendship. Unfortunately, he passed away on March 30, 2005. Also to Reuben Hersh who taught me patiently how to write English properly (although I still need further improvements) and for writing the Foreword. Words cannot express my gratitude to Cristina and Cathy for their friendship and help. Without their constant encouragement, this book project may have been aborted. The Contest owes its smooth operation to Dan Cosper, Kathy Hall, Shirley Harty, Jenison Klinger, Moira Robertson, and Ron Schrader. Their contributions are heartily appreciated. I express my gratitude to Dirk Bridwell, Peter Espen, Elizabeth Frank, Chuck Mader, Ron Stewart, Craig Lewis, Dann Brewer, and my son, Shin-Yi, and daughter-in-law, Dorothy, for their help whenever I encounter computer problems, and Susan Pinter for her assistance with illustrations. It is a pleasure to thank Mina Yamashita, Lisa Pacheco, Maya Allen-Gallegos, Glenda Madden, and Luther Wilson of UNM Press. Their devotion to the production of a quality book is beyond my expectation. Last but not least, I am deeply grateful to my nieces, Lisa Chen and Ingrid Hahn Chen; niece-in-law, Terry Huang; daughter-in-law, Rosemary; and my three sons, Shin-Yi, Shin-Jen, and Shin-Hong. All of them read early drafts and commented from various perspectives, which resulted in considerable improvements.

Finally, a word about my name. Its Mandarin pronunciation is Liang-shin, but I prefer the Taiwanese one: Liong-shin.

—L.-s. H.

PART ONE
Problems

Chapter 1.1 Number Theory and Algebra

1. A rectangular chocolate bar consists of 30 small rectangular chocolate pieces arranged in five rows of six. If you 'fold' it along a 'valley', then you will break it into two pieces. What is the minimum number of 'foldings' needed to break the chocolate bar into 30 small pieces? You are not allowed to arrange two or more pieces together in one 'folding'; i.e., for each 'folding', you are only permitted to break one piece into two.

2. There are 50 teams in the State Basketball Tournament. One loss eliminates a team and there are no ties. How many games must be played to decide which team wins the championship?

3. Andrew, Beau, Carolyn, and Darcey played chess. Every player played two games with each of the other players. For each game, the winner got 2 points, while the loser got 0 points. In case of a tie, each player got 1 point. At the end, Andrew got 6 points, Beau got 3 points, and Carolyn got 8 points. How many points did Darcey get?

4. The distance between two points, A and B, is 18 *meters*, and there are poles every 60 *cm* between them. If we want to change the distance between each pair of neighboring poles from 60 *cm* to 75 *cm*, then what is the minimum number of poles we have to remove? (Notes: 1 *meter* = 100 *cm*, and originally there are 31 poles.)

5. Hester has a pocketful of pennies, nickels, dimes and quarters. (There is at least one coin of each value.)

   If she has a total of 19 coins worth 93 cents, how many dimes does she have?

6. A merchant wanted to sell his radios for \$149 each, but no one would buy any. He decided to cut the price, and he offered the radios at a reduced price. At the end of the sale, he told a friend: "The reduced price was more than half the original price and I sold all the radios for a grand total of \$1978." If the reduced price was a whole number of dollars, how many radios did he sell?

7. Find all prime numbers p such that $1999p + 1$ is a perfect square.

2

8. A survey showed that among a certain set of 50 people only 3 spoke all three languages, English, French, and Spanish, while 18 spoke exactly two of these languages and 25 spoke just one of these languages. A total of 35 spoke English. Also, a total of 9 spoke French but each of these also spoke English or Spanish or both.

   (a) How many of the 50 people spoke Spanish?

   (b) Were there any people in the survey who did not speak any of these three languages? If so, how many were there?

9. Figure 9 is an example of a magic square of order 4. The sum of the four integers in each row, column, or diagonal is always 34. So we say the magic sum of this magic square is 34. Suppose there is a magic square of order 7, with integers from 1 through 49. What would be the magic sum?

16	3	2	13
5	10	11	8
9	6	7	12
4	15	14	1

Figure 9

10. In a magic square of addition, the sum of the integers in each row, column and diagonal is the same. For example, in Figure 10(a), the magic sum is 15. Fill in the blanks in Figure 10(b) with positive integers to make it a magic square of multiplication; i.e., complete Figure 10(b) with positive integers so that the product of the three positive integers in each row, column, and diagonal becomes the same.

4	9	2
3	5	7
8	1	6

Figure 10(a)

3		
4		
	1	

Figure 10(b)

11. Figure 11(a) and Figure 11(b) are examples of magic squares of order 3 and 5, respectively. (In each case, the sum of the integers in each row, column, or diagonal is the same.)

19	93	32
61	48	35
64	3	77

Figure 11(a)

8	21	17	5	14
2	15	9	23	16
24	18	1	12	10
11	7	25	19	3
20	4	13	6	22

Figure 11(b)

Observe that the number 48 at the center of the magic square of order 3 is the average of all the entries in that magic square:

$$48 = \frac{1}{9}(19 + 93 + 32 + 61 + 48 + 35 + 64 + 3 + 77).$$

Is this always true of every magic square of order 3? That is, if Figure 11(c) is a magic square, then is it necessary that

$$E = \frac{1}{9}(A + B + C + D + E + F + G + H + I)\,?$$

A	B	C
D	E	F
G	H	I

Figure 11(c)

Justify your assertion.

12. Express 1988 as a sum of two or more consecutive positive integers in as many different ways as you can.

13. Of the following 10 numbers, only one is a perfect square. Which one?

(*a*)	8344572651	(*b*)	7955896032
(*c*)	1695032253	(*d*)	4906358264
(*e*)	1782570645	(*f*)	5729581636
(*g*)	3213046377	(*h*)	2032918848
(*i*)	2973562479	(*j*)	3567659100

(This may look tedious and dull; try a shortcut.)

14. Prove or disprove: There exist (infinitely many) sets of 10 consecutive integers with the property that the last two digits of their squares (in increasing order) are

$$81,\ 64,\ 49,\ 36,\ 25,\ 16,\ 09,\ 04,\ 01,\ 00.$$

If you think the proposition is correct, exhibit two such sets with smallest positive integers. If you think the proposition is wrong, give your reason.

15. Can an integer with 2 or more digits, all of which are either 1, 3, 5, 7, or 9 (for example, 1991, 17, 731591375179, 753), be a perfect square? Justify your assertion.

16. On reviewing last year's New Mexico Mathematics Contest problems, Veronica said: "A perfect square whose last digit is 6 must have an odd digit right before the last digit." Paige said: "I was thinking the converse: a perfect square whose next-to-the-last digit is odd must have 6 as the last digit."

Prove or disprove each of their assertions.

17. (a) Find the remainder when 1993^2 is divided by 9.
    (b) For some integer exponent n, we have

$$1993^n = 15777A7325840B,$$

where two missing digits are replaced by A and B. Find A, B and n.

18. Find the integer n such that $n^5 = 4984209207$.

19. During the show-and-tell hour at school, Kira asked everyone in her class to choose his or her own non-palindromic three-digit number (say, 928). Then, reverse the order of its digits (829), and subtract the small one from the large one ($928-829 = 99$). Supply a zero in front or back of the difference, if necessary, to make it a three-digit number (099, say). Finally, reverse the order of the digits again (990), and add (instead of subtract) the two numbers this time ($099 + 990$). Now Kira asked, "Does everyone get 1089?", and her classmates responded with a thunderous "Yes!"

Explain the reason behind Kira's trick.

20. Two cousins, Amy and Kira, enjoy playing mathematics games together. One day, Amy said to Kira: "Choose any two numbers, then let the third number be the sum of the first and the second, the fourth number the sum of the second and the third, and so on. That is, from the third number on, each one is the sum of the two numbers immediately preceding it. For example, supposing your first two numbers are 2 and 1, respectively, then you have the sequence:

$$2, 1, 3, 4, 7, 11, 18, 29, 47, 76, \cdots.$$

Now just tell me the seventh number in your sequence, and I'll guess the sum of the first 10 numbers in your sequence." "O.K.," said Kira. "Now I choose the first two numbers, and I find the seventh number in my sequence to be 99." Instantly, Amy found the sum of the first 10 numbers in Kira's sequence correctly. What is Amy's answer?

21. During the holidays, Sierra visited Aunt Lisa in California.

"Aunt Lisa, show me some of your mathematics tricks," said Sierra.

"O.K., if you wish," replied Aunt Lisa. "Choose any positive integer, say 314159, and rearrange its digits in any way you want and get, say 193415; then take (the absolute value of) their difference, in our case we have $314159 - 193415 = 120744$. Now, you hide away one of the nonzero digits in the difference, and just tell me the rest of the digits in any order, say 4, 2, 1, 0, 4. Then I'll guess the nonzero digit you hid away; in our example, it's 7."

"Oh, that's a cool mathematics game to play with kindergarten kids, but I see the trick behind it right away," said Sierra.

"Really? O.K., let me test you," said Aunt Lisa with an air of doubt. "I choose a number, scramble its digits and find the digits in the difference to be 6, 5, 8, 0, 3, in some order, with one nonzero digit hidden away. Now tell me the nonzero digit I hid away."

Instantly, Sierra found Aunt Lisa's hidden digit correctly.

(a) What is Aunt Lisa's hidden digit?

(b) Explain the reason behind this trick.

22. One summer day, Nicky visited Uncle Paul in Los Alamos.

"Uncle Paul, show me some of your mathematics tricks," said Nicky.

"O.K., if you wish," replied Uncle Paul. "Choose a number between 0 and 100, and just tell me the remainders when it is divided by 3, 5 and 7, respectively. Then I'll guess your secret number."

"Let me see, my secret number gives the remainders 2, 4 and 5, when it is divided by 3, 5 and 7, respectively," said Nicky.

"Abracadabra! Your secret number is 89," replied Uncle Paul within seconds.

"Right! But how did you find it so fast?" asked Nicky.

"It is very simple, Nicky," said Uncle Paul. "All you need is to remember the magic triple $\{70, 21, 15\}$. Your secret number gives the remainder 2 when divided by 3, so I multiply 70 by 2 to get 140. Similarly, I multiply 21 by 4 to get 84, and 15 by 5 to get 75. Then the sum $140 + 84 + 75 = 299$ has the property that it gives the same remainders as your secret number when it is divided by 3, 5, and 7. But 299 is not between 0 and 100, so I reduce it by a multiple of 105 ($= 3 \times 5 \times 7$), to get 89 ($= 299 - 2 \times 105$), which is your secret number."

"That's neat, Uncle Paul. I'll think it over tonight. Thank you and good-bye."

Early next morning, Nicky went to see Uncle Paul again.

"Uncle Paul, choose a number between 0 and 1000, and just tell me the remainders when it is divided by 7, 11 and 13, respectively. Then I'll guess your secret number."

"O.K., my secret number gives remainders 1, 9 and 4 when it is divided by 7, 11 and 13, respectively."

Within a minute, Nicky was able to find Uncle Paul's secret number correctly.

(a) What is Uncle Paul's secret number?

(b) What is Nicky's magic triple?

23. Observe that $\{49, 50, 51, 52\}$ is an example of a set of four consecutive positive integers that are divisible, respectively, by 7, 5, 3, and 2. Find the four smallest consecutive positive integers that are divisible, respectively, by 2, 3, 5, and 7.

24. Find positive integers x, y, and z satisfying

$$x + \frac{y}{19} + \frac{z}{97} = \frac{1997}{19 \times 97}.$$

25. (a) Find the positive integers p and q (which have no common divisor other than 1) such that

$$\frac{p}{q} = 0.\dot{1}8\dot{5} = 0.185185185\cdots.$$

(b) Find the positive integers r and s (which have no common divisor other than 1) such that

$$\frac{r}{s} = 0.1\dot{4}8\dot{6} = 0.1486486486\cdots.$$

26. (a) Find all the possible denominators that give repeating decimals of minimum length 3 and without non-repeating parts; i.e., find all the positive integers q with the property that, given a positive integer p which has no common divisor (other than 1) with q, there exist 3 digits a, b, c (not all equal) and a nonnegative integer n such that

$$\frac{p}{q} = n.\dot{a}b\dot{c} = n.abcabcabc\cdots.$$

(b) Fill in all the blanks, and show that the solution is unique.

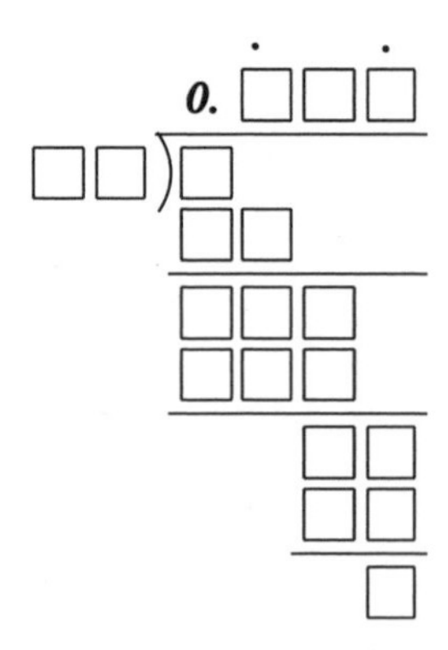

Figure 26

27. The senior class at Rocky Mountain High School has 100 students and they want to elect five of their members to represent them at the State Fair. Each senior votes for two classmates, and the five students getting the most votes will be sent to the Fair. What is the least number of votes which a student would need to be assured of being elected?

28. Find the minimum value of all the numbers in the sequence

$$\sqrt{\frac{7}{6}}+\sqrt{\frac{96}{7}},\ \sqrt{\frac{8}{6}}+\sqrt{\frac{96}{8}},\ \sqrt{\frac{9}{6}}+\sqrt{\frac{96}{9}},\ \cdots,\ \sqrt{\frac{95}{6}}+\sqrt{\frac{96}{95}}.$$

29. Find all real numbers x satisfying the inequality

$$\frac{x-1}{x-3} \geq \frac{x-2}{x-4}.$$

30. Find the integer n such that

$$n \leq \frac{1}{\sqrt{1}}+\frac{1}{\sqrt{2}}+\frac{1}{\sqrt{3}}+\cdots+\frac{1}{\sqrt{119}}+\frac{1}{\sqrt{120}} < n+1.$$

(**Hint:** $\dfrac{1}{\sqrt{k}+\sqrt{?}} < \dfrac{1}{2\sqrt{k}} < \dfrac{1}{\sqrt{?}+\sqrt{k}}$; or if you wish you may use calculus.)

31. Find the positive integer k satisfying

$$\frac{1}{k+1} < \left(\sqrt{3}-\sqrt{2}\right)^4 < \frac{1}{k}.$$

32. (a) Find positive integers a and b satisfying

$$\tan\frac{3\pi}{8} = a+\sqrt{b}.$$

(b) Show that

$$\left(\tan\frac{3\pi}{8}\right)^n + (-1)^n\left(\cot\frac{3\pi}{8}\right)^n$$

is an even integer for every positive integer n.

(c) For each positive integer n, let k_n be a positive integer such that

$$k_n < \left(\tan\frac{3\pi}{8}\right)^n < k_n + 1.$$

Show that k_n and n are of opposite parity; i.e., one is even and the other is odd.

(d) Find k_5.

33. Observe that, for $k = 40$,

$$2k + 1 = 2 \cdot 40 + 1 = 9^2, \quad 3k + 1 = 3 \cdot 40 + 1 = 11^2.$$

We want to show that there exist infinitely many positive integers k for which $2k + 1$ and $3k + 1$ are both perfect squares. Suppose

$$2k + 1 = m^2, \quad 3k + 1 = n^2,$$

where m and n are positive integers. Define an integer K by the magic formula:

$$K = 11m^2 + 20mn + 9n^2.$$

Then $2K+1$ and $3K+1$ are both perfect squares, because for suitable positive integers p, q, r, s (and m, n as above),

$$2K + 1 = (pm + qn)^2, \quad 3K + 1 = (rm + sn)^2.$$

It follows that for any positive integer k for which $2k + 1$ and $3k + 1$ are both perfect squares, our magic formula gives a bigger integer K with the same property. Hence, there exist infinitely many positive integers with the desired property. Determine the coefficients p, q, r, s.

34. Let a, b, c and k be nonzero real numbers satisfying the relations

$$k = \frac{a}{b+c} = \frac{b}{c+a} = \frac{c}{a+b}.$$

Find all the possible common values k of these fractions.

35. Find three distinct odd positive integers a, b, and c so that

$$\frac{1}{3} = \frac{1}{a} + \frac{1}{b} + \frac{1}{c}.$$

36. Find rational numbers x, y, u, v satisfying

$$\begin{aligned} z &= x - \sqrt{2}, & w &= 5 + y\sqrt{2}, \\ z + w &= u + \sqrt{2}, & zw &= 11 + v\sqrt{2}. \end{aligned}$$

37. (a) Find positive integers u and v satisfying

$$\sqrt{18 - 2\sqrt{65}} = \sqrt{u} - \sqrt{v}.$$

(b) Find positive integers x and y satisfying

$$\sqrt{14 + 3\sqrt{3 + 2\sqrt{5 - 12\sqrt{3 - 2\sqrt{2}}}}} = x + \sqrt{y}.$$

38. Find the positive integers a, b, c and p satisfying

$$\sqrt{94 + a\sqrt{p}} = b + c\sqrt{p},$$

where p is a prime number and $c > 1$.

39. Assume coefficients p, q, r exist such that, for every positive integer n,

$$\frac{1^4 + 2^4 + 3^4 + \cdots + (n-1)^4 + n^4}{1^2 + 2^2 + 3^2 + \cdots + (n-1)^2 + n^2} = pn^2 + qn + r.$$

Determine the coefficients p, q, r.

40. Using the patterns of the equalities:

$$\begin{aligned} 3^2 + 4^2 &= 5^2, \\ 5^2 + 12^2 &= 13^2, \\ 7^2 + 24^2 &= 25^2, \\ 9^2 + 40^2 &= 41^2, \\ 11^2 + 60^2 &= 61^2, \end{aligned}$$

find positive integers x and y such that $13^2 + x^2 = y^2$.

41. Suppose a, b, c are the lengths of the sides of a triangle. If $a^2 + ab + b^2 = c^2$, then the largest angle of the triangle is $\frac{2\pi}{3}$. Observe that

$$\begin{array}{rcrcrcrcr} 5^2 & + & 5 & \times & 3 & + & 3^2 & = & 7^2, \\ 7^2 & + & 7 & \times & 8 & + & 8^2 & = & 13^2, \\ 9^2 & + & 9 & \times & 15 & + & 15^2 & = & 21^2, \\ 11^2 & + & 11 & \times & 24 & + & 24^2 & = & 31^2, \\ 13^2 & + & 13 & \times & 35 & + & 35^2 & = & 43^2. \end{array}$$

Find positive integers x and y such that $17^2 + 17x + x^2 = y^2$.

42. Observe that

$$\begin{aligned} 3^2 + 4^2 &= 5^2, \\ 3^2 + 4^2 + 12^2 &= 13^2, \\ 3^2 + 4^2 + 12^2 + 84^2 &= 85^2. \end{aligned}$$

Find a pair of positive integers x and y such that

$$3^2 + 4^2 + 12^2 + 84^2 + x^2 = y^2.$$

(It is not necessary to find all such pairs.)

43. In solving a quadratic equation, Newton misread the coefficient of the first degree term and obtained 3 and -4 as solutions, while Leibnitz misread the constant term and obtained -1 and 5 as solutions. Find the correct solutions.

44. (a) Find integers a and b such that $\dfrac{4+3\sqrt{3}}{2+\sqrt{3}}$ is a root of the quadratic polynomial $x^2 + ax + b$.

    (b) Suppose $f(x) = x^4 + 2x^3 - 10x^2 + 4x - 10$, and $f\left(\dfrac{4+3\sqrt{3}}{2+\sqrt{3}}\right) = c\sqrt{3} + d$, where c and d are integers. Find the values of c and d.

45. Two quadratic equations

$$1997x^2 + 1998x + 1 = 0 \quad \text{and} \quad x^2 + 1998x + 1997 = 0$$

have a root in common. Find the product of the roots that are not in common.

46. Suppose the two quadratic equations

$$x^2 - 5x + k = 0 \quad \text{and} \quad x^2 - 9x + 3k = 0$$

have a nonzero root in common. What is the value of k?

47. Suppose a cubic polynomial $x^3 + px^2 + qx + 72$ is divisible by both $x^2 + ax + b$ and $x^2 + bx + a$ (where a, b, p, q are constants and $a \neq b$).

    Find the roots of the cubic polynomial.

48. (a) Solve for x: $|x|x - 5x - 6 = 0$.

    (b) Find all real numbers x satisfying the inequality

$$|x|x - 5x - 6 > 0.$$

49. Suppose the quartic polynomial

$$x^4 + 2x^3 - 23x^2 + px + q$$

has a pair of double roots (say, a, a, b, b). Find p and q.

50. Find all integer values of x for which

$$x^2 - 7x - 4$$

is a perfect square.

51. Let $y = x + \frac{1}{x}$. Then $x^2 + \frac{1}{x^2} = \left(x + \frac{1}{x}\right)^2 - 2 = y^2 - 2$.

   (a) Express $x^4 + \frac{1}{x^4}$ in terms of y.

   (b) Express $x^5 + \frac{1}{x^5}$ in terms of y.

52. (a) Find all possible values of y, where $y = x + \frac{1}{x}$, and x satisfies the equation

$$x^4 + 2x^3 - 22x^2 + 2x + 1 = 0.$$

   (b) Solve the quartic polynomial

$$x^4 + 2x^3 - 22x^2 + 2x + 1 = 0.$$

53. Suppose α and β are the two roots of the quadratic equation $3x^2 + 4x + 5 = 0$. Find the value of

$$\frac{\alpha}{\beta} + \frac{\beta}{\alpha}.$$

54. Let α, β, γ be the three roots of the cubic equation $x^3 - 2x + 3 = 0$.

   (a) Find the value of $\alpha^2 + \beta^2 + \gamma^2$.

   (b) Find the value of $\alpha^3 + \beta^3 + \gamma^3$.

55. Expressions

$$s_1 = a + b + c, \quad s_2 = bc + ca + ab, \quad s_3 = abc$$

are known as basic symmetric forms of a, b, c. Every symmetric form of a, b, c can be expressed (uniquely) in terms of s_1, s_2, s_3. For example,

$$a^2 + b^2 + c^2 = s_1^2 - 2s_2.$$

   (a) Express $(b + c)(c + a)(a + b)$ in terms of s_1, s_2, s_3.

   (b) Express $a^3 + b^3 + c^3$ in terms of s_1, s_2, s_3.

56. Suppose $u + v = 3$, $u^2 + v^2 = 13$.

   (a) Find the value of uv.

   (b) Find the value of $u^3 + v^3$.

57. Suppose $\sin\theta - \cos\theta = \frac{\sqrt{5}}{2}$.

   (a) Find the value of $\sin\theta \cdot \cos\theta$.

   (b) Find the value of $\sin^3\theta - \cos^3\theta$.

58. Suppose all the roots of the quartic polynomial

$$x^4 + 2x^3 + ax^2 + bx + 15,$$

where a and b are integers, are rational.

(a) Find all the roots of the quartic polynomial.

(b) Find the values of a and b.

59. Let α and β be the roots of the quadratic equation

$$(x-2)(x-3) + (x-3)(x+1) + (x+1)(x-2) = 0.$$

Evaluate

$$\frac{1}{(\alpha+1)(\beta+1)} + \frac{1}{(\alpha-2)(\beta-2)} + \frac{1}{(\alpha-3)(\beta-3)}.$$

60. (a) Show that there exists an angle φ such that

$$3\sin\theta + 4\cos\theta = 5\sin(\theta+\varphi) \quad \text{for all} \quad \theta.$$

(b) Find the necessary and sufficient condition on the constant k such that the trigonometric equation

$$k\sin\theta - 2\cos\theta = \sqrt{7}$$

has a solution.

61. (a) Find the constants h and k such that the following becomes an identity

$$\frac{1}{x^2-1} = \frac{h}{x-1} + \frac{k}{x+1}.$$

(b) Evaluate

$$\frac{1}{2^2-1} + \frac{1}{3^2-1} + \frac{1}{4^2-1} + \cdots + \frac{1}{10^2-1}.$$

62. Suppose $a_1, a_2, \cdots, a_n, \cdots$ is a sequence of numbers with the property that

$$a_1 + a_2 + \cdots + a_n = \frac{n(n+1)(n+2)}{6} \quad \text{for all} \quad n = 1, 2, 3, \cdots.$$

(a) Find a_{19}.

(b) Evaluate

$$\frac{1}{a_1} + \frac{1}{a_2} + \cdots + \frac{1}{a_{94}} + \frac{1}{a_{95}}.$$

63. Define

$$S(x) = \frac{2^x - 2^{-x}}{2}, \quad C(x) = \frac{2^x + 2^{-x}}{2}, \quad T(x) = \frac{S(x)}{C(x)}.$$

(a) Express $S(x+y)$ in terms of $S(x)$, $S(y)$, $C(x)$, and $C(y)$.

(b) Express $C(x+y)$ in terms of $S(x)$, $S(y)$, $C(x)$, and $C(y)$.

(c) Express $T(x+y)$ in terms of $T(x)$ and $T(y)$.

64. A function g is said to be even, while a function h is said to be odd, if

$$g(-x) = g(x), \qquad h(-x) = -h(x) \qquad \text{for all } x.$$

For example, $g(x) = 3 + 5x^2$ is even, while $h(x) = 2x - x^3$ is odd.

(a) Given a function

$$f(x) = \frac{1}{1 - x + x^2},$$

find a pair of functions g and h, where g is even and h is odd, such that

$$f(x) = g(x) + h(x) \quad \text{for all real numbers } x.$$

(b) Is such a decomposition of f unique?

65. Observe that if

$$g(x) = x + \frac{1}{x}, \quad h(x) = \frac{x+2}{2x+1},$$

then, for all $x > 0$, we have

$$g\left(\frac{1}{x}\right) = g(x) > 0, \quad h\left(\frac{1}{x}\right) = \frac{1}{h(x)}.$$

(a) Given a function $f(x) = \dfrac{x}{x^2+3}$, find a pair of functions, g and h, where, for all $x > 0$, we have

$$g\left(\frac{1}{x}\right) = g(x) > 0, \quad h\left(\frac{1}{x}\right) = \frac{1}{h(x)} \quad \text{and} \quad f(x) = g(x) \cdot h(x).$$

(b) Is such a decomposition of f unique?

66. Suppose f is an odd function defined for all real x; i.e.,

$$f(-x) = -f(x) \quad \text{for all real number } x;$$

yet by shifting the function by 1, it becomes an even function; i.e., $g(x) = f(x+1)$ is even, namely

$$g(-x) = g(x) \quad \text{for all real number } x.$$

In terms of f, this means

$$f(-x+1) = f(x+1) \quad \text{for all real number } x.$$

(a) Show that f is a periodic function, and find its period; i.e., find the smallest positive number p such that

$$f(x+p) = f(x) \quad \text{for all real number } x.$$

(b) Give an example of such a function.

67. (a) Find, if any, a function f satisfying

$$f\left(\frac{x}{x-1}\right) = x \quad (x \neq 1).$$

(b) Find, if any, a nonconstant function g satisfying

$$g\left(\frac{x}{x-1}\right) = g(x) \quad (x \neq 1).$$

How many such functions are there?

68. Find a function f satisfying the functional equation

$$f\left(\frac{1}{1-x}\right) + f\left(\frac{x-1}{x}\right) = \frac{x}{x-1} \quad \text{for all } x \neq 0,\ 1.$$

Is such a function unique?

69. Observe that function

$$f(x) = x + \frac{1}{x}$$

has the properties that

$$\begin{aligned} f\left(\frac{1}{x}\right) &= f(x) \quad \text{for } x > 0; \quad \text{and} \\ f(x) - f(1) &= x + \frac{1}{x} - 2 = \frac{(x-1)^2}{x} \end{aligned}$$

has a double root at $x = 1$.

Suppose g is a nonconstant rational function such that

$$g\left(\frac{1}{x}\right) = g(x) \quad \text{for } x > 0.$$

State and prove a proposition about the multiplicity (order) of the root of

$$g(x) - g(1) = 0$$

at $x = 1$.

70. Given a polynomial

$$p(x) = \sum_{k=0}^{2003} a_k x^k = a_0 + a_1 x + a_2 x^2 + \cdots + a_{2003} x^{2003},$$

express the following polynomials in terms of p:

(a) $f_0(x) = \sum_{k=0}^{667} a_{3k} x^{3k} = a_0 + a_3 x^3 + a_6 x^6 + \cdots + a_{2001} x^{2001};$

(b) $f_1(x) = \sum_{k=0}^{667} a_{3k+1}x^{3k+1} = a_1 + a_4x^4 + a_7x^7 + \cdots + a_{2002}x^{2002};$

(c) $g(x) = \sum_{k=0}^{500} a_{4k}x^{4k} = a_0 + a_4x^4 + a_8x^8 + \cdots + a_{2000}x^{2000}.$

71. Does there exist a finite sequence of numbers, real or complex,

$$\{a_k\,;\; k = 1,\, 2,\, \cdots,\, n\}$$

satisfying the following inequalities simultaneously?

1^0 $\quad |a_k| < 1 \quad$ (for all $k = 1, 2, \cdots, n$).

2^0 $\quad \left|\sum_{k=1}^{n} a_k\right| < \left|\sum_{k=1}^{n} a_k^2\right| < \left|\sum_{k=1}^{n} a_k^3\right| < \left|\sum_{k=1}^{n} a_k^4\right| < \left|\sum_{k=1}^{n} a_k^5\right|.$

72. Express an arbitrary positive integer n as ordered sums of positive integers in 2^{n-1} ways. For example, if $n = 5$, the 16 ordered sums are listed in the left column below:

5	$2\ (= 2)$
$4+1$	$2\times 1\ (= 2)$
$1+4$	$1\times 2\ (= 2)$
$3+2$	$2\times 3\ (= 6)$
$2+3$	$3\times 2\ (= 6)$
$3+1+1$	$2\times 1\times 1\ (= 2)$
$1+3+1$	$1\times 2\times 1\ (= 2)$
$1+1+3$	$1\times 1\times 2\ (= 2)$
$2+2+1$	$3\times 3\times 1\ (= 9)$
$2+1+2$	$3\times 1\times 3\ (= 9)$
$1+2+2$	$1\times 3\times 3\ (= 9)$
$2+1+1+1$	$3\times 1\times 1\times 1\ (= 3)$
$1+2+1+1$	$1\times 3\times 1\times 1\ (= 3)$
$1+1+2+1$	$1\times 1\times 3\times 1\ (= 3)$
$1+1+1+2$	$1\times 1\times 1\times 3\ (= 3)$
$1+1+1+1+1$	$1\times 1\times 1\times 1\times 1\,(= 1)$

The entries in the right column are obtained from the corresponding entries in the left column by changing (a) additions to multiplications and (b) all the integers greater than or equal to 3 to 2, (c) 2 to 3, (d) 1 to 1 (i.e., keeping 1 unchanged).
Now add all the products in the right column. For $n = 5$, we obtain

$$2+2+2+6+6+2+2+2+9+9+9+3+3+3+3+1 = 64\ (= 8^2).$$

Prove or disprove: For every positive integer n, the sum of all the products in the right column is always a perfect square.

Chapter 1.2 Geometry and Combinatorics

101. Determine the sum of the twelve angles marked in Figure 101.

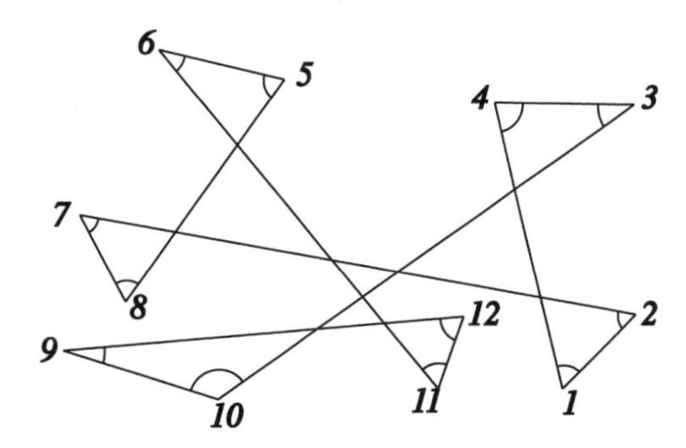

Figure 101

102. A polygon that does not cross itself and whose vertices all lie on lattice points is called a simple lattice polygon. (Lattice points are points whose coordinates are integers.) A theorem of Pick asserts that there exist coefficients a, b, and c with the property that if H is a simple lattice polygon with p lattice points in its interior and q lattice points on its boundary (for example, in Figure 102, $p = 14$, and $q = 33$), then

$$\text{Area of } H = ap + bq + c.$$

In other words, it is possible to find the area of an arbitrary simple lattice polygon by just counting the lattice points. Assuming such coefficients exist, determine what these coefficients a, b, c must be.

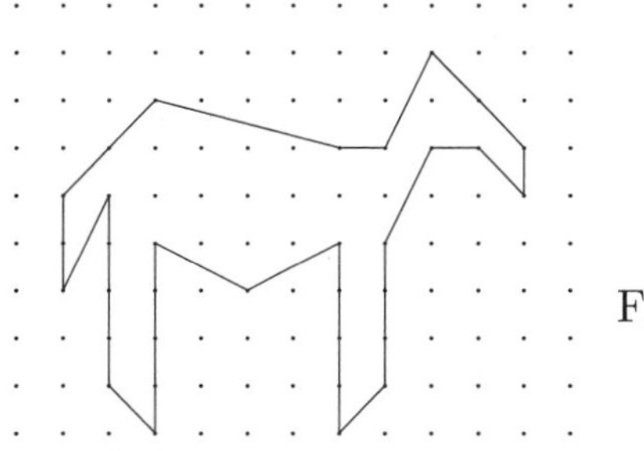

Figure 102

103. Does there exist an equilateral triangle (in the (x, y)-plane) all three of whose vertices are at lattice points? If your answer is Yes, give an example of such an equilateral triangle by indicating the coordinates of the 3 vertices; if your answer is No, state your reason.

104. (a) Does there exist a triangle similar to a 3-4-5 right triangle having all its vertices at lattice points, and exactly one of its sides is parallel to the coordinate axes?

(b) Does there exist a triangle similar to a 3-4-5 right triangle having all its vertices at lattice points, yet none of its sides is parallel to the coordinate axes?

(c) What if the 3-4-5 right triangle is replaced by an arbitrary Pythagorean right triangle?

105. There are 30 lattice points, 5 by 6, as in Figure 105. Of all the squares with their 4 vertices at these lattice points,

(a) how many have a pair of horizontal sides?

(b) how many do not have sides that are horizontal?

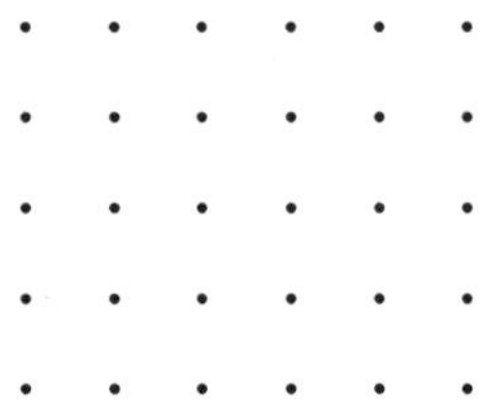

Fig 105

106. Arrange three squares of the same size as in Figure 106.

Find the sum of the two angles $\angle DAE$ and $\angle DBE$.

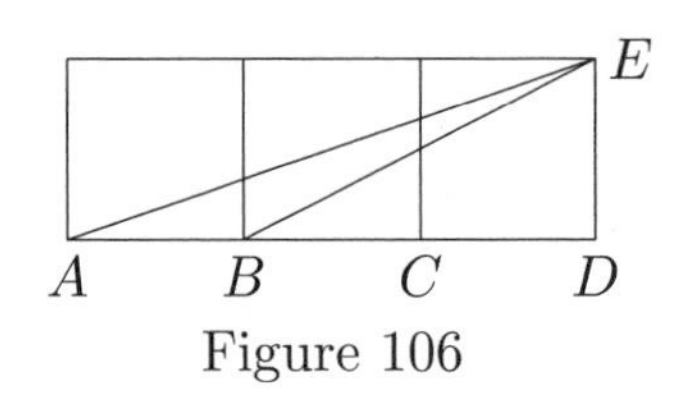

Figure 106

107. Let P be a point inside the square $ABCD$ such that $\triangle PAB$ is an equilateral triangle. Find $\angle CPD$.

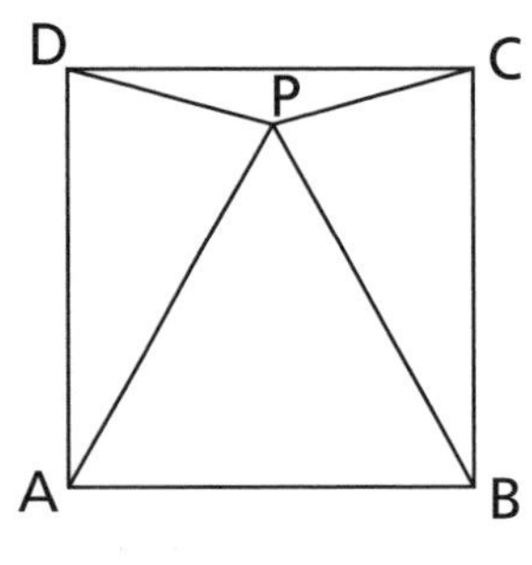

Figure 107

108. In Figure 108, the three line segments AB, EF, CD are perpendicular to the line BD, and the three points A, E, D are collinear (i.e., on a straight line), as are the points B, E, C and B, F, D. Suppose $\overline{AB} = 10\,cm$, $\overline{CD} = 15\,cm$.

Find length $\overline{EF}$.

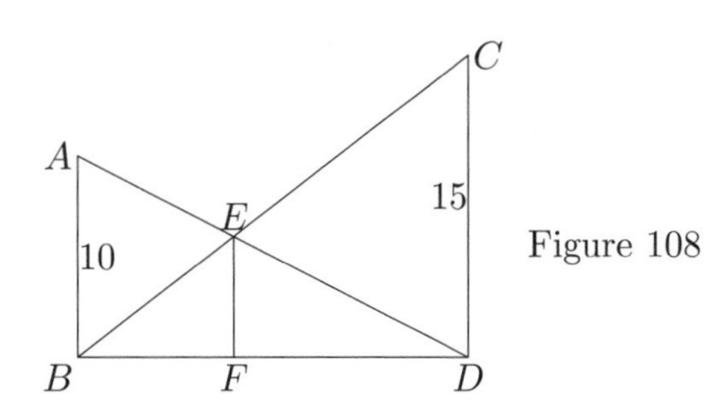

Figure 108

109. We have a billiard table measuring 15 units by 8 units as shown in Figure 109. We hit a billiard ball, which is at the point P, 3 units from the lower left corner of the table. It then hits three sides of the table and returns to its starting point. The angle at which the ball strikes the side of the table is the same as the angle at which it leaves the side of the table. How far did the ball travel?

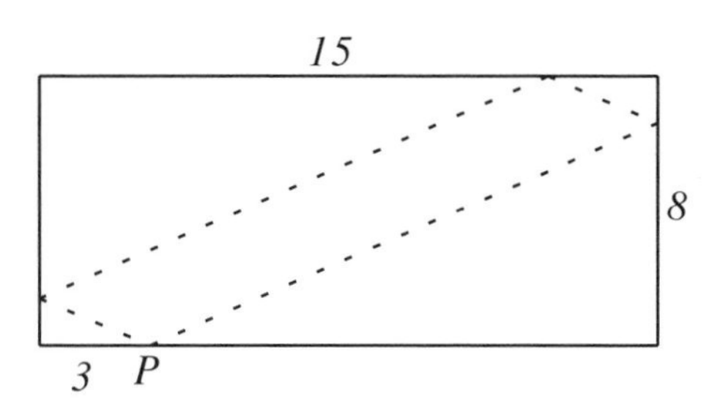

Figure 109

110. Lines from the vertices of a parallelogram to the midpoints of the sides are drawn as shown in Figure 110, forming a smaller parallelogram of area 7 cm^2 in the center. What is the area of the original parallelogram?

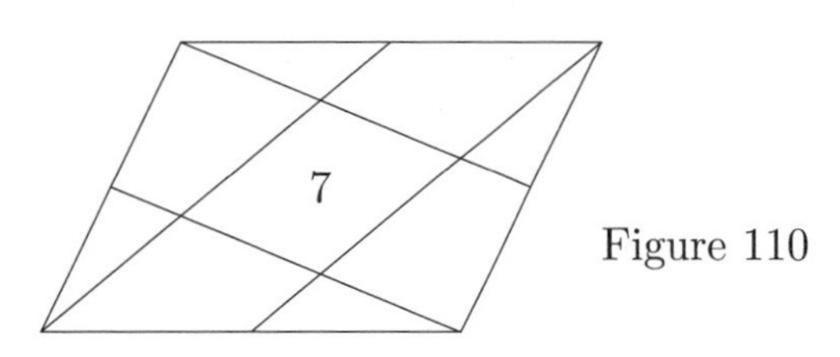

Figure 110

111. Find the volume of a rectangular box whose three faces meeting at a vertex have areas 12, 18 and 24 cm^2.

112. Suppose E and F are the midpoints of sides BC and CD, respectively, of parallelogram $ABCD$. If the area of $\triangle AEF$ is 12 cm^2, what is the area of the parallelogram

$ABCD$?

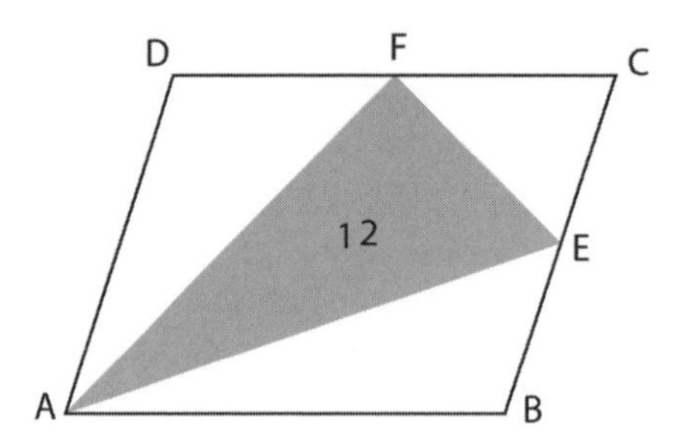

Figure 112

113. Suppose $\triangle AEF$ is inscribed in a rectangle $ABCD$ as shown in Figure 113. If the area of $\triangle AEF$ is 25 cm^2, and

$$\overline{BE} = 4\ cm, \quad \overline{DF} = 6\ cm,$$

find the area of the rectangle $ABCD$.

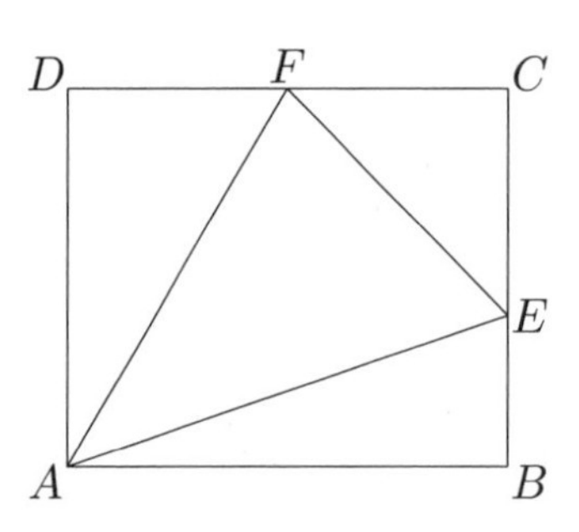

Figure 113

114. Suppose a quadrangle $EFGH$ is inscribed in a rectangle $ABCD$ as shown in Figure 114. Let G' and H' be points on AB and BC, respectively, such that GG' is parallel to DA and HH' is parallel to AB. If the area of the quadrangle $EFGH$ is 32 cm^2, and

$$\overline{EG'} = 3\,cm, \quad \overline{FH'} = 5\,cm,$$

find the area of the rectangle $ABCD$.

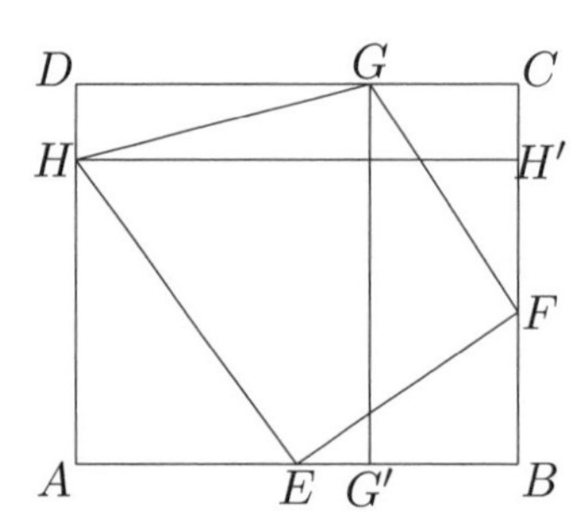

Figure 114

115. Let $\triangle ABC$ be a right triangle. Choose two points D and E on the hypotenuse AB such that

$$\overline{AD} = \overline{AC}, \quad \overline{BE} = \overline{BC}.$$

Suppose $\overline{AB} = 29\ cm$ and $\overline{DE} = 12\ cm$.
What are the lengths of the legs BC and AC?

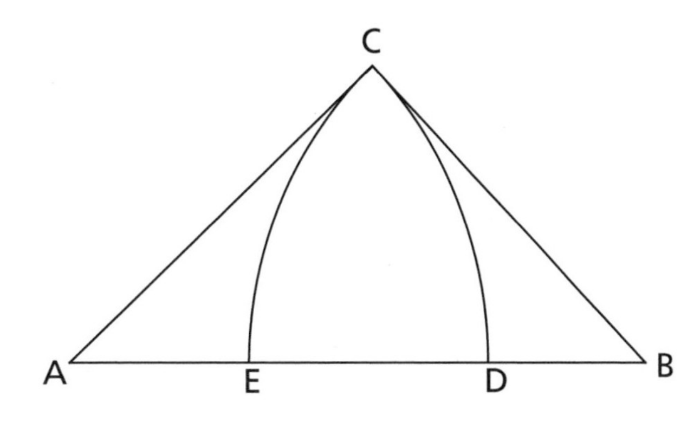

Figure 115

116. (a) Find the area of an isosceles triangle whose sides have lengths 16 cm, 17 cm, 17 cm.

(b) Suppose there is an obtuse (one angle is larger than $\frac{\pi}{2}$) isosceles triangle whose two equal sides have length 17 cm and whose area is the same as that of the acute triangle given in Part (a). Find the length of the third side.

117. In $\triangle ABC$, let X and Y be the feet of perpendiculars from vertex A to the bisectors of (interior) angles B and C, respectively. Suppose the lengths of AB, AC and XY are 10, 7 and 3 cm, respectively. Find the length of side BC.

118. Given a triangle ABC with area 1 cm^2, extend its sides BC, CA, AB to points D, E, F, respectively, so that $\overline{BD} = 2\overline{BC}$, $\overline{CE} = 3\overline{CA}$, $\overline{AF} = 4\overline{AB}$. Find the area of $\triangle DEF$.

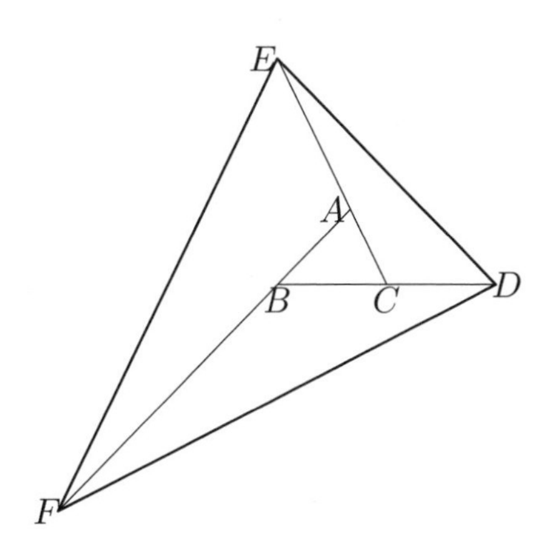

Figure 118

119. On the sides of $\triangle ABC$, choose points P, Q, R, so that

$$\overline{BP} = \overline{PC}, \quad \overline{CQ} = 2\overline{QA}, \quad \overline{AR} = 2\overline{RB}.$$

Joining the points P, Q, R divides $\triangle ABC$ into four small triangles. If the areas of these four triangles are four consecutive integers (with cm^2 as the unit), what is the area of $\triangle ABC$?

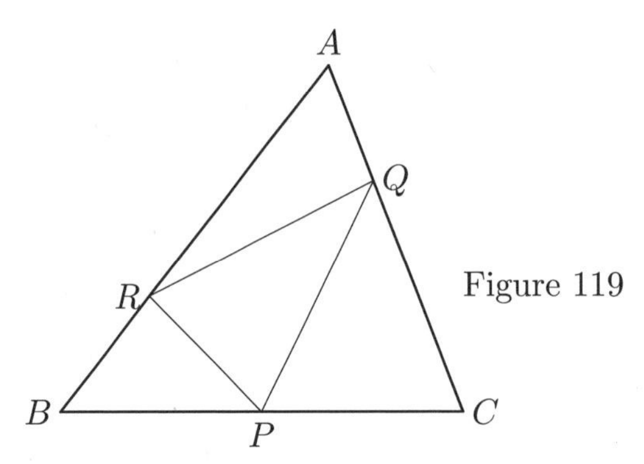

Figure 119

120. In $\triangle ABC$, suppose

$$\overline{AB} = 5\,cm, \quad \overline{AC} = 7\,cm, \quad \angle ABC = \frac{\pi}{3}.$$

(a) Find the length of the side BC.

(b) Find the area of $\triangle ABC$.

121. Suppose square $DEFG$ is inscribed in $\triangle ABC$, with vertices D, E on side AB, and F, G on sides BC, CA, respectively. Given $\overline{AB} = 28\,cm$ and the length of a side of square $DEFG$ is $12\,cm$, determine, if possible, the area of $\triangle ABC$.

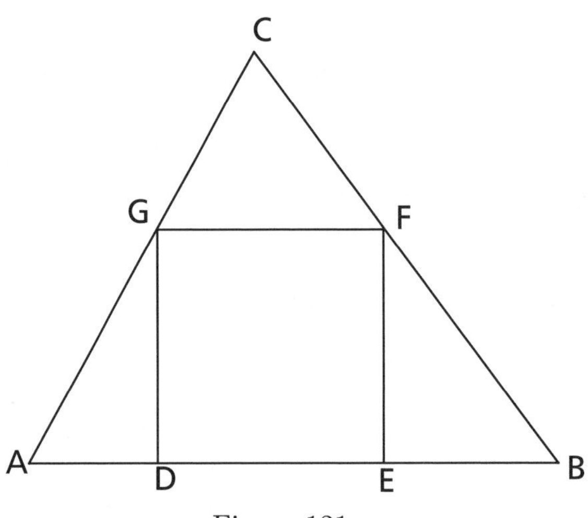

Figure 121

122. The circumradius of a triangle is the radius of the circle, called the circumcircle, that passes through the three vertices of the triangle. The inradius of a triangle is the

radius of the circle, called the incircle, that is inside the triangle and is tangent to the three sides of the triangle.

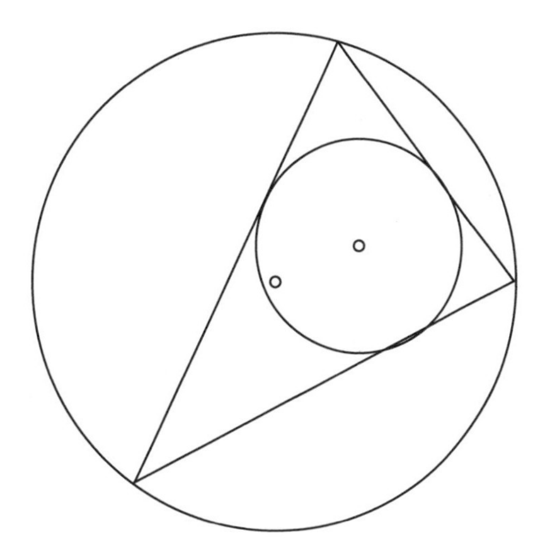

Figure 122

(a) Find the circumradius R of the 3-4-5 right triangle.

(b) Find the inradius r of the 3-4-5 right triangle.

123. (a) Express the area of a right triangle in terms of its circumradius and inradius.

(b) Express the lengths of the three sides of a right triangle in terms of its circumradius and inradius.

124. Let P be a point in the plane of an equilateral triangle ABC such that

$$\triangle PBC, \quad \triangle PCA, \quad \triangle PAB$$

are all isosceles. (A triangle is isosceles if at least two of its sides are of the same length; it is equilateral if all three of its sides are of the same length.)

How many such points P are there?

125. (a) What is the maximum number of acute angles a convex polygon can have? (A set is convex if it includes, with each pair of its points, the entire line segment joining them; e.g., every triangle is convex and so is a circle, but neither a star nor a cross is convex. An angle is acute if it is less than $\frac{\pi}{2}$.)

(b) If a convex polygon has exactly 5 obtuse angles, what is the maximum possible number of its sides? (An angle is obtuse if it is more than $\frac{\pi}{2}$.)

126. Suppose there are eight points on a circle, equally spaced. (See Figure 126.)

(a) How many triangles are there having all their vertices at three of these eight points?

(b) How many of these triangles are right triangles?

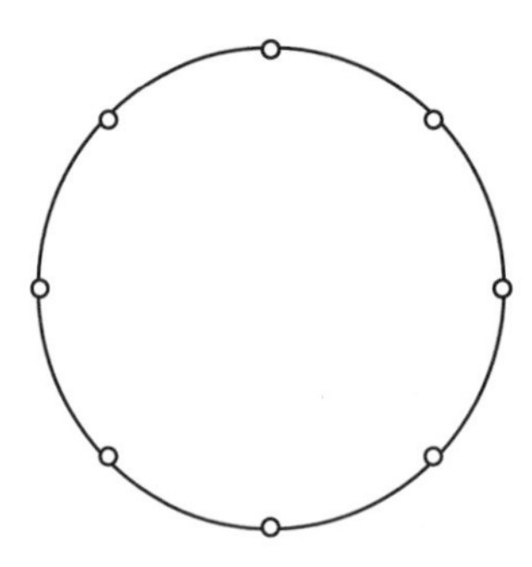

Figure 126

127. Suppose there are 12 points on a circle, equally spaced.

 (a) Of all the triangles having all their vertices at three of these 12 points, how many have at least one 60^0 angle?

 (b) How many of these triangles are of the type 30^0-60^0-90^0?

128. Of all the triangles having all their vertices at three of the vertices of a given convex n-gon (i.e., a convex polygon with n sides), $7n$ of them share no side with the polygon. Find the value of n.

129. (a) A cube is sliced by a plane so that the cross-section is an equilateral triangle. What is the maximum possible length of a side of such an equilateral triangle (assuming the length of each edge of the cube is 1)? How many distinct slices are there of this type?

 (b) A cube is sliced by a plane so that the cross-section is a regular hexagon. What is the length of a side of such a regular hexagon (assuming the length of each edge of the cube is 1)? How many distinct slices are there of this type?

130. We have a box (a rectangular parallelepiped) whose height, width and length are 3, 4 and 5 cm, respectively.

 Suppose the box is sliced by a plane so that the cross-section is a square.

 (a) What is the maximum possible length of a side of the square? How many distinct slices are there of this type?

 (b) What is the minimum possible length of a side of the square? How many distinct slices are there of this type?

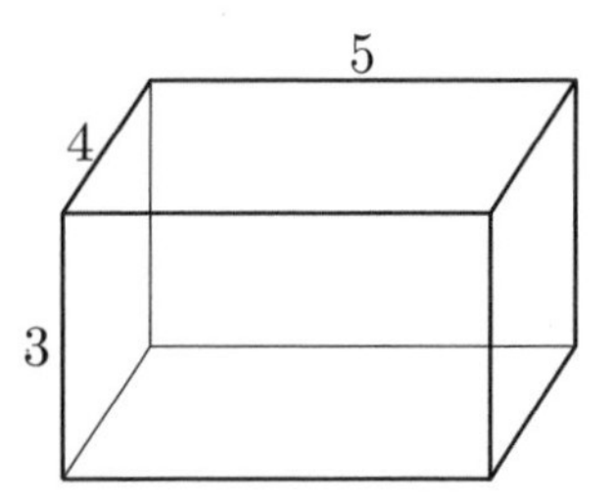

Figure 130

131. (a) Of all the triangles having all their vertices at the vertices of a given cube, how many of them are right triangles?

  (b) Describe the remaining triangles, if any, that are not right triangles. How many triangles are there of this type?

132. Suppose AC is one of the sides of an equilateral triangle having all of its vertices at the vertices of the cube obtained by folding the figure on the left. Indicate the other two sides (using the labels in the figure on the left). Is the answer unique?

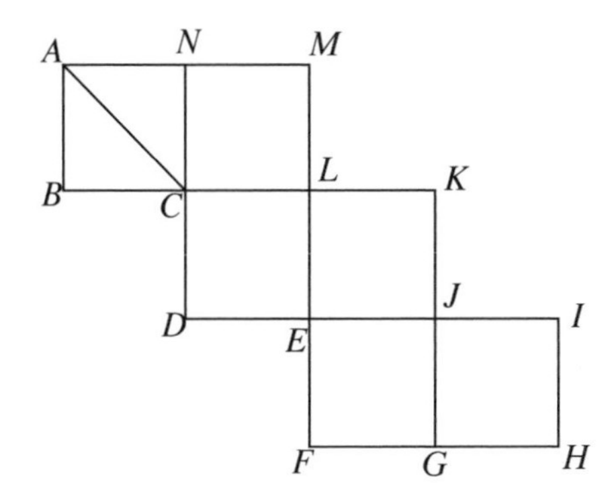

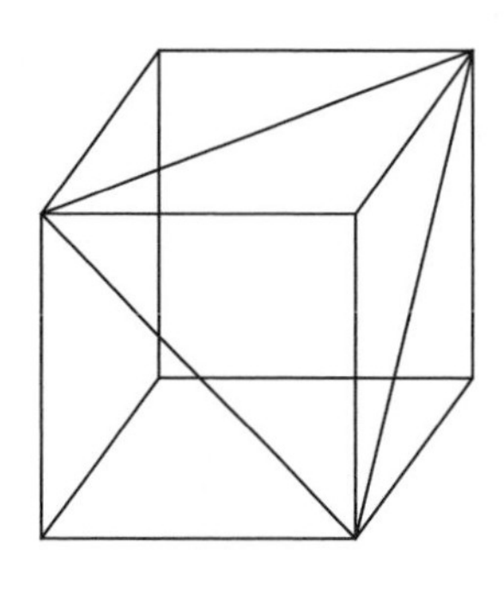

Figure 132

133. We want to paint all five regions in Figure 133 such that no neighboring regions are painted in the same color.

  (a) If we have four different colors, how many ways are there to paint the regions?

  (b) What if we have five different colors?

In each case, it is not required to use all the colors.

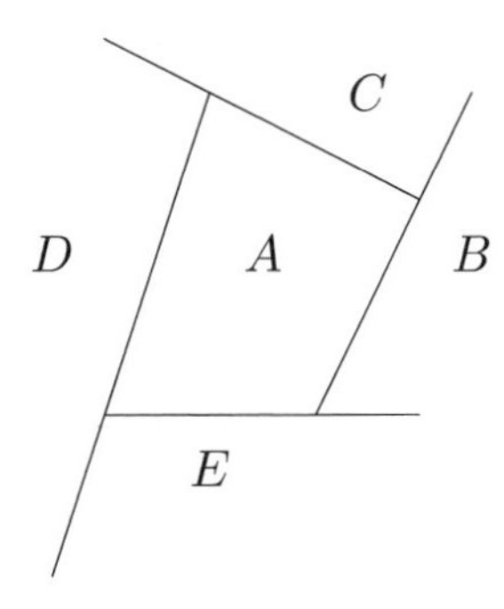

Figure 133

134. (a) In how many ways can the faces of a cube be painted by six different colors if no two faces are to be painted in the same color?

  (b) What if the cube in Part (a) is replaced by a box (i.e., a rectangular parallelepiped), none of whose faces is a square?

135. A bug is at point A on the right circular cone shown in Figure 135. If the bug crawls to the point P on the surface of the cone by the shortest route, how far does it crawl? (A right circular cone is made by cutting a wedge out of a circular disc and gluing the cut edges together.) The lengths of the line segments shown in Figure 135 are the following (where AB is a diameter of the base):

$$\overline{OA} = \overline{OB} = \overline{AB} = 15\,cm, \quad \overline{OP} = 8\,cm.$$

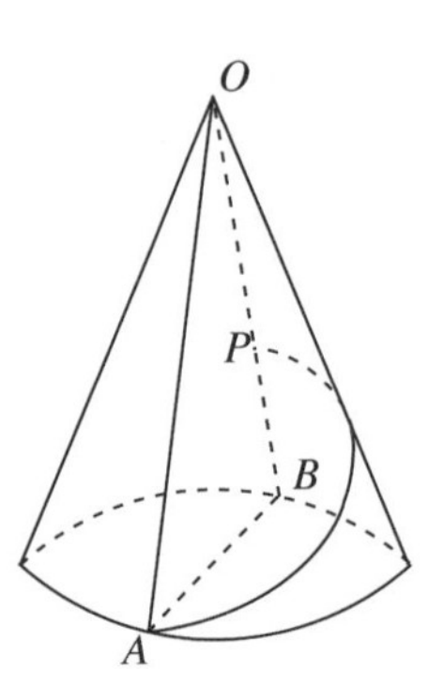

Figure 135

136. Figure (a) and Figure (b) are examples of quadrangles whose four corners can be folded to meet at a point without either overlapping or forming a gap. Of the next 6 quadrangles (Figures (c) - (h)), determine which one(s) can be folded in the same manner.

Figure (c) $\overline{AB} = \overline{CD}$, $\overline{AD} = \overline{BC}$, $\overline{AB} \neq \overline{AD}$, $\angle A = \pi/2$.

Figure (d) $\begin{cases} KN \perp KL. \\ K, L, M, N \text{ are midpoints of the respective sides.} \end{cases}$

Figure (e) $\begin{cases} KM \perp LN, \overline{KM} \neq \overline{LN}. \\ K, L, M, N \text{ are midpoints of the respective sides.} \end{cases}$

Figure (f) $\overline{AB} = \overline{AD}$, $\overline{BC} = \overline{CD}$.

Figure (g) $\overset{\frown}{AB} + \overset{\frown}{CD} = \overset{\frown}{BC} + \overset{\frown}{AD}$.

Figure (h) $AC \perp BD$.

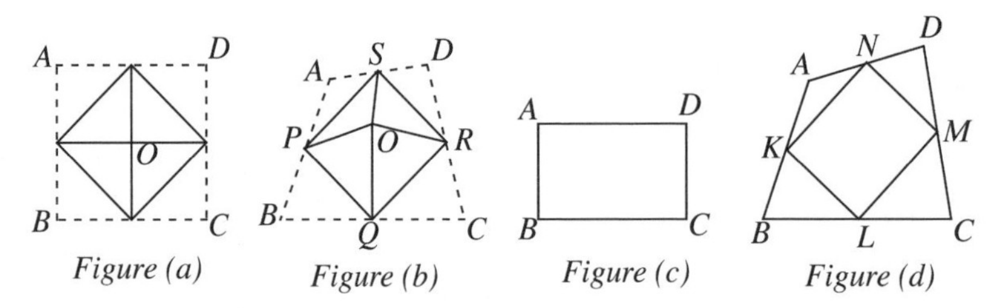

Figure 136

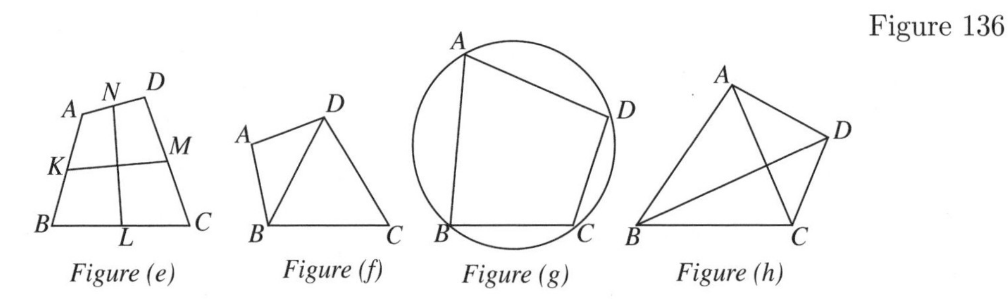

137. A square sheet of paper $ABCD$ is folded as shown in Figure 137 with D falling on E, which is on BC, with A falling on F, and EF intersecting AB at G.

Prove that, regardless of the position of E on the side BC, the perimeter of $\triangle EBG$ is a fixed length, and express this length in terms of the length ℓ of a side of the square $ABCD$.

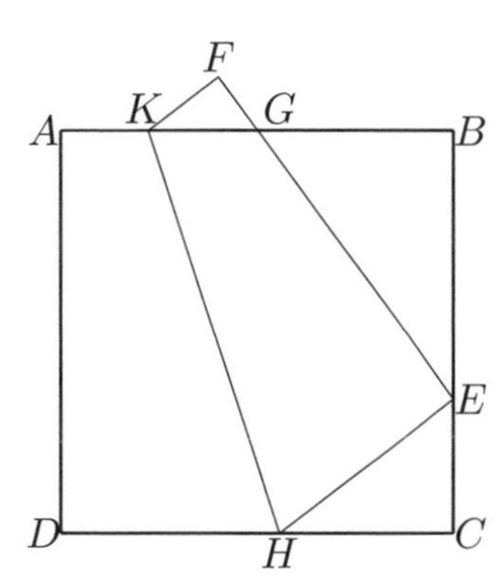

Figure 137

138. A square sheet of paper $ABCD$ is folded along GH as shown in Figure 138, with A falling on E, which is on BC, and D falling on F.

Suppose

$$\overline{AB} = 18\ cm, \quad \overline{BE} = 6\ cm.$$

Find the area of the trapezoid $EFGH$.

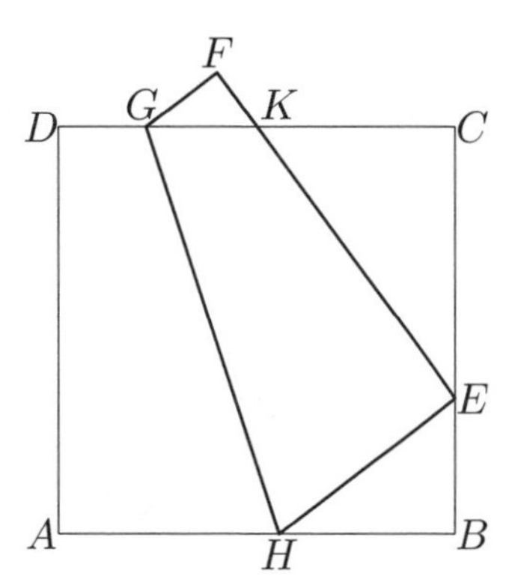

Figure 138

139. Suppose a square is projected orthogonally to a line (the square and the line are coplanar) such that the images of the two diagonals have lengths 7 and 3 cm, respectively; i.e.,

$$\overline{A'C'} = 7, \quad \overline{B'D'} = 3$$

in Figure 139. What is the length of a side of the square?

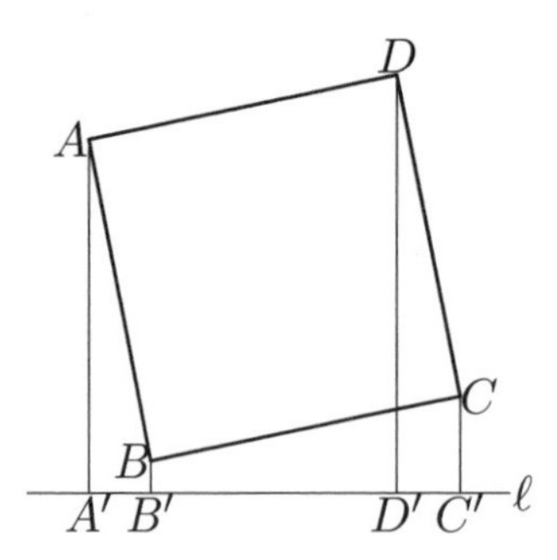

Figure 139

140. Suppose a rectangle is projected orthogonally to a line (the rectangle and the line are coplanar) such that the images of the two diagonals have lengths 7 and 5 cm, respectively.

    (a) What is the minimum possible area of the rectangle?

    (b) Find the dimensions of the rectangle that give the minimum area.

141. In Figure 141, let $ABCD$ be a convex quadrangle. Points E, F trisect side AB; G, H trisect side BC; I, J trisect side CD; K, L trisect side DA. Joining points E and J, F and I, G and L, H and K divide quadrangle $ABCD$ into 9 small quadgrangles. Prove or disprove: The area of quadrangle $PQRS$ at the center is $1/9$ of that of the original quadrangle $ABCD$.

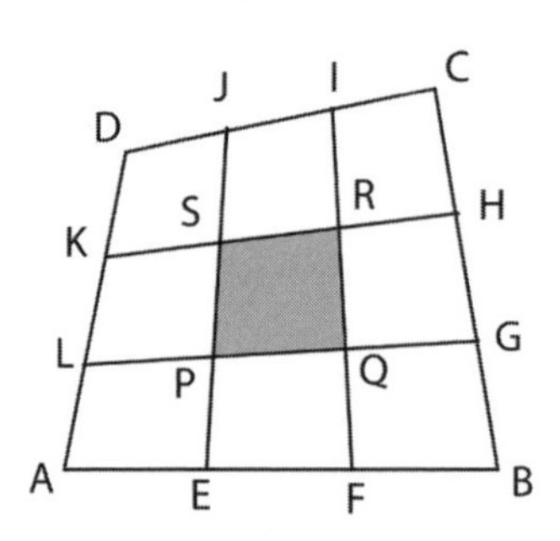

Figure 141

142. In a quadrangle $ABCD$, suppose

$$\begin{aligned} \overline{AB} &= 5\ cm, & \overline{CD} &= 8\ cm, \\ \angle ABC &= 70^0, & \angle BCD &= 50^0, \end{aligned}$$

and P, Q, R, S are the midpoints of BC, CA, AD, DB, respectively. Find the area of the parallelogram $PQRS$.

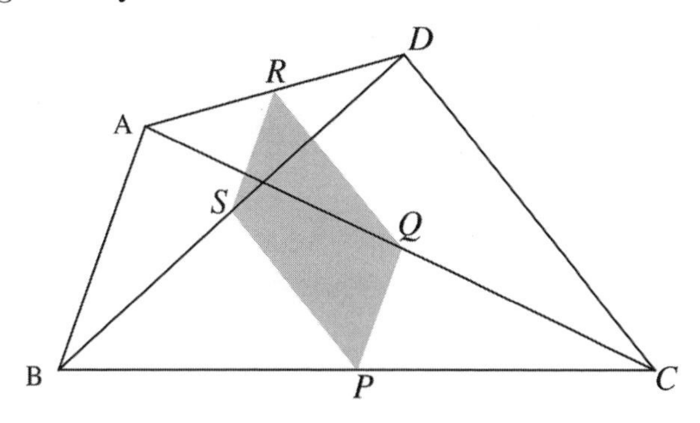

Figure 142

143. (a) In the obtuse triangle ABC, the angle at vertex C is $\frac{\pi}{4}$. Let D be the foot of the perpendicular from C to the extension of AB. If the length of AD is 6 cm and the length of BD is 1 cm, find the area of the triangle ABC.

(b) Explain geometrically why the answer is not unique.

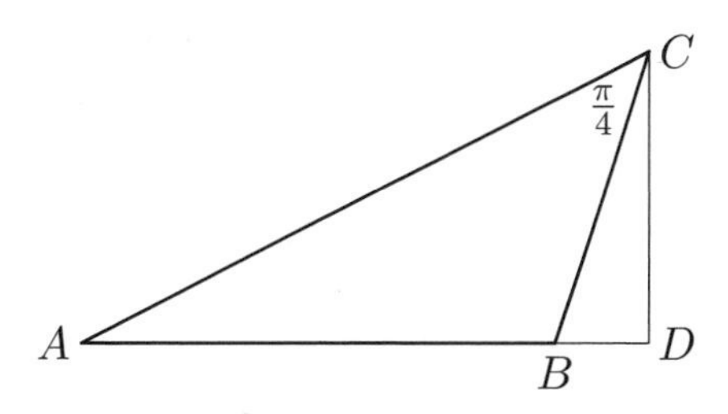

Figure 143

144. On the sides AB, AC of $\triangle ABC$, choose two points D and E, respectively, in such a way that $\triangle ADE$ and the quadrangle $DBCE$ have the same areas and the same perimeters. Given

$$\overline{BC} = 13\,cm, \quad \overline{CA} = 15\,cm, \quad \overline{AB} = 16\,cm,$$

find the lengths $\overline{AD}$ and $\overline{AE}$. Is the answer unique?

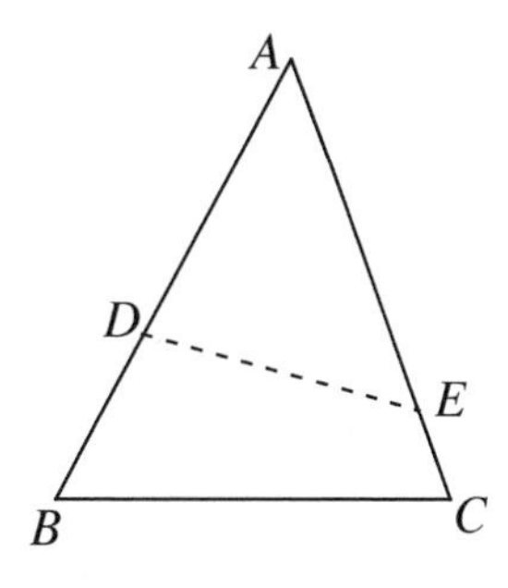

Figure 144

145. Through a point P inside the triangle ABC, three lines parallel to the sides are drawn so that $\triangle ABC$ is divided into three triangles and three parallelograms. If the triangles have areas 1 cm^2, 4 cm^2, and 9 cm^2, what is the area of the original $\triangle ABC$?

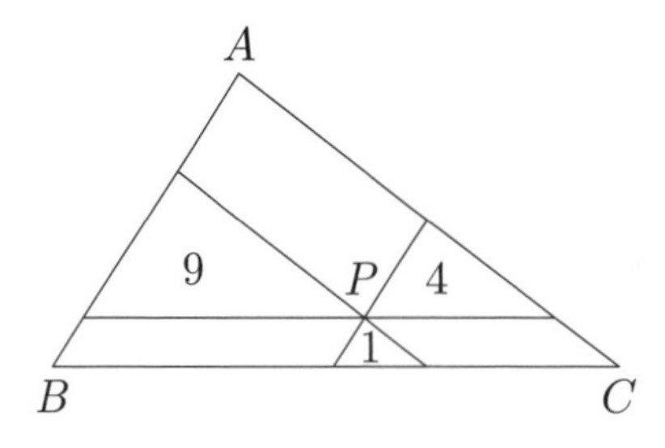

Figure 145

146. Through a point P inside the triangle ABC, three lines parallel to the sides are drawn so that $\triangle ABC$ is divided into three triangles and three parallelograms. If the parallelograms have areas 6 cm^2, 10 cm^2, and 15 cm^2, what is the area of the original $\triangle ABC$?

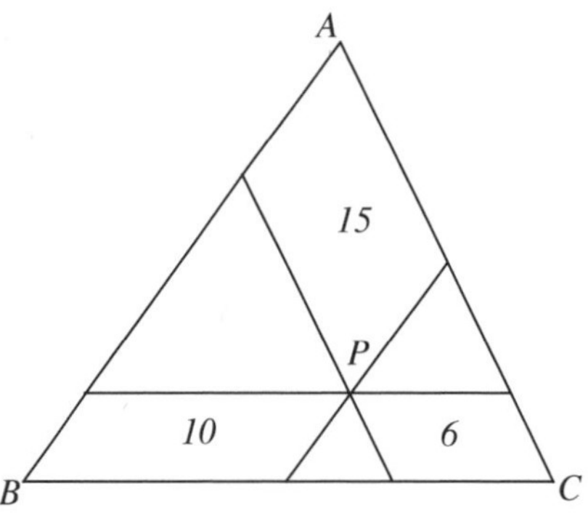

Figure 146

147. Suppose the diagonals AC and BD of a trapezoid $ABCD$ intersect at P. If the areas of $\triangle PAB$ and $\triangle PCD$ are 36 cm^2 and 25 cm^2, respectively, what is the area of the trapezoid $ABCD$? (Note that $AB \parallel CD$.)

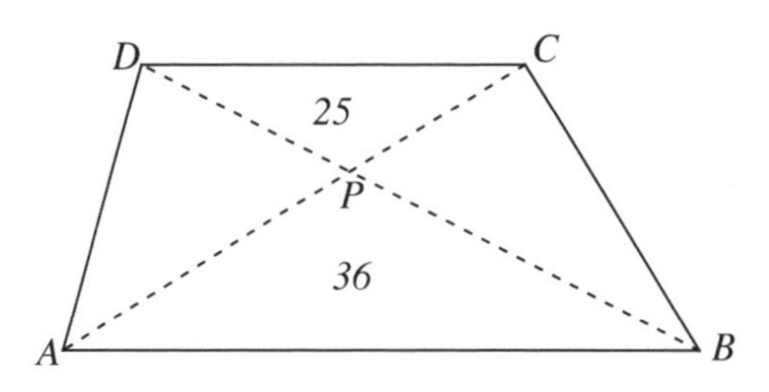

Figure 147

148. Suppose the diagonals AC and BD of a convex quadrangle $ABCD$ intersect at P, and the areas of $\triangle PAB$ and $\triangle PCD$ are 16 cm^2 and 25 cm^2, respectively.

What is the minimum possible area of the quadrangle $ABCD$?

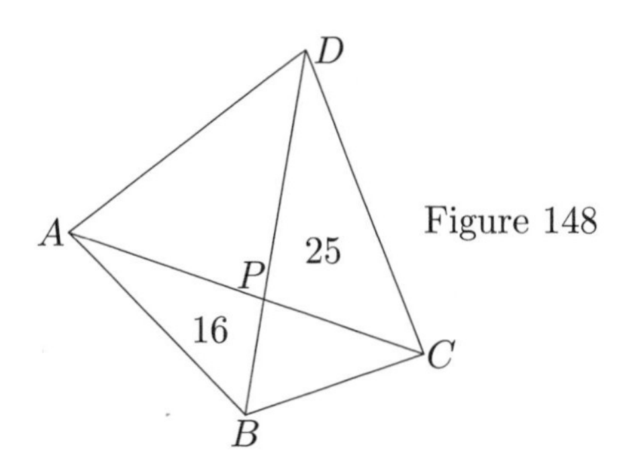

Figure 148

149. Suppose AB is the common chord of two circles O and O', C and D are points on the circle O and O', respectively, such that BC is tangent to the circle O', while BD is tangent to the circle O. Suppose

$$\overline{AC} = 3, \quad \overline{AD} = 4;$$

find the length of the common chord AB.

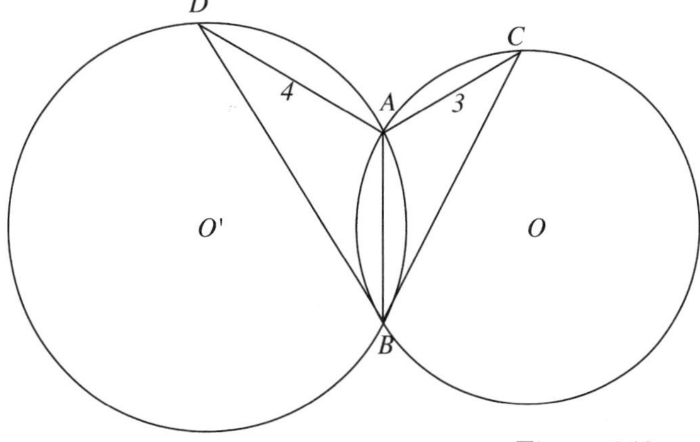

Figure 149

150. Let D and E be the feet of the perpendiculars from vertices B and C of $\triangle ABC$ to the opposite sides AC and AB, respectively, and M the midpoint of the side BC. Find the condition on $\triangle ABC$ such that $\triangle MDE$ is equilateral.

151. Suppose P is a point on a line segment AB, and equilateral triangles PAC, PBD are drawn on the same side of the line AB.

   (a) Describe the locus of the intersection Q of AD and BC as P moves from A to B (along the line segment AB).
   (b) Find the length of the locus of Q, assuming $\overline{AB} = 9$.

152. In Figure 152, points A, B, G, S are on a circle whose center is at M, and O is a point on the extension of the diameter AB. Moreover, OG is tangent to the circle and both GH and SM are perpendicular to the diameter AB.
Suppose $\overline{OA} = a$ and $\overline{OB} = b$ $(0 < a < b)$. Express the inequalities

$$\overline{OH} < \overline{OG} < \overline{OM} < \overline{OS}$$

in terms of a and b.

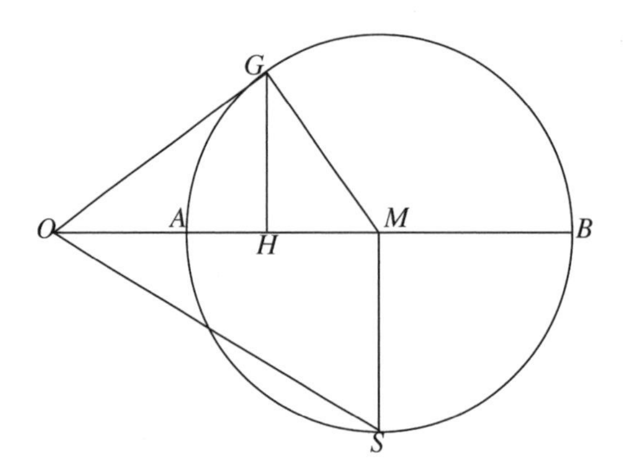

Figure 152

153. In Figure 153, the side AB of a cyclic quadrangle $ABCD$ is a diameter of the circumcircle, and

$$\overline{BC} = 7, \quad \overline{CD} = \overline{DA} = 3.$$

What is the length of the diameter AB?

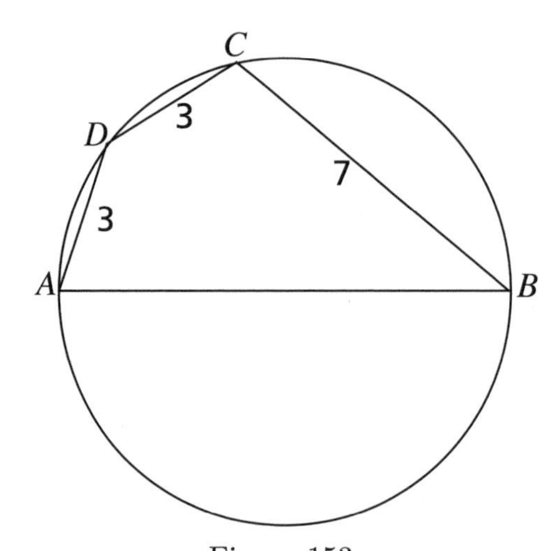

Figure 153

154. Given six fixed points on a circle, label them as A, B, C, D, E, F in any order. Let P, Q, R denote the intersection points (of the extensions if necessary) of the 'opposite sides' AB and DE, BC and EF, CD and FA, respectively (assuming they all intersect). A theorem of Pascal, which Pascal discovered when he was 16 years old, says that points P, Q, and R lie on a line. Let us call this line a *Pascal Line* of the 'hexagon' $ABCDEF$. Two examples of Pascal lines are shown in Figure 154. Notice that these hexagons may be self-intersecting as in the figure on the right. Since the six points can be labelled in any order, by changing the order of the labels, we may

obtain different Pascal lines. What is the maximum possible number of Pascal lines (for the same six points)?

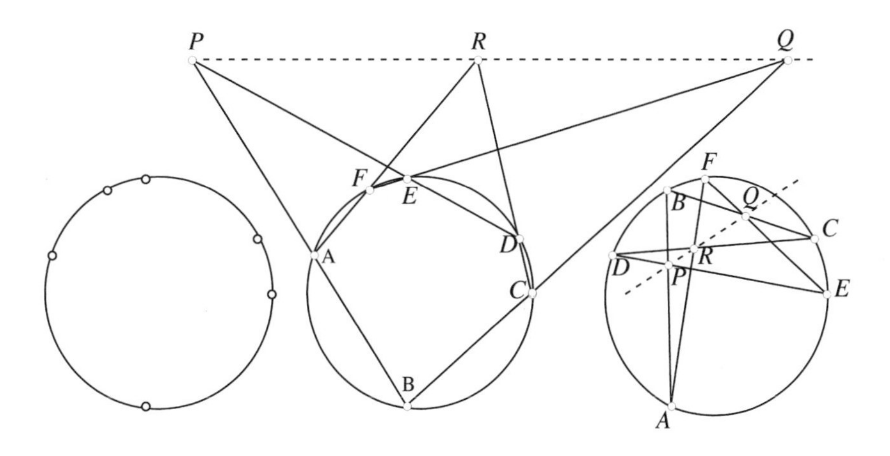

Figure 154

155. Let P be a point in the plane of the rectangle $ABCD$. If $\overline{BP} = 5$, $\overline{CP} = 11$, $\overline{DP} = 10$, find the length $\overline{AP}$.

156. Let P be a point inside an equilateral triangle ABC. The distances $\overline{PA}$, $\overline{PB}$, $\overline{PC}$ are 1, 2, $\sqrt{3}$ cm, respectively.

  (a) Find the angles around the point P; i.e., find $\angle BPC$, $\angle CPA$ and $\angle APB$.

  (b) Find the length of side BC of the equilateral triangle ABC.

157. Suppose P is a point inside a square $ABCD$ such that

$$\overline{PA} = 3, \quad \overline{PB} = 2\sqrt{2}, \quad \overline{PC} = 5.$$

  (a) Find $\angle APB$.

  (b) Find the length of side AB of the square $ABCD$.

158. (a) Find integers a and b satisfying

$$\tan\frac{\pi}{8} = \sqrt{a} - b.$$

Hint: In Figure 158, $\angle ABC = \frac{\pi}{2}$, $\quad \overline{AB} = \overline{BC}$, $\quad \overline{AC} = \overline{CD}$.

$$\therefore \angle ADB = \frac{\pi}{8}, \quad \tan\frac{\pi}{8} = \frac{\overline{AB}}{\overline{BD}}.$$

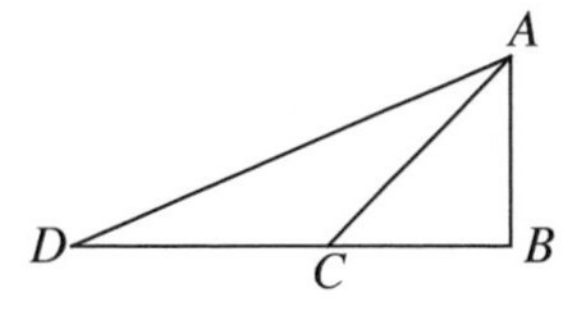

Figure 158

(b) Find integers c and d satisfying

$$\tan\frac{\pi}{12} = c - \sqrt{d}.$$

(c) Find integers p, q, r, s satisfying

$$\tan\frac{\pi}{24} = (\sqrt{p} - \sqrt{q})(\sqrt{r} - s).$$

159. Note that the set of all the points (x, y) in the plane satisfying the inequality

$$|x| + |y| \leq 1$$

is a square whose area is 2 (square units).

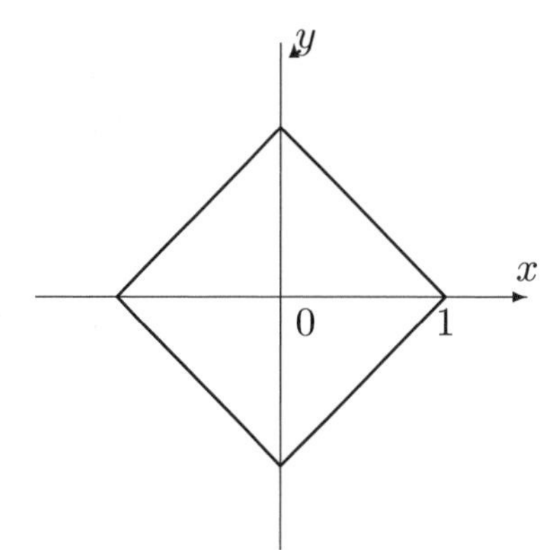

Figure 159

The set of all the points (x, y) in the plane satisfying the inequality

$$\Big||x| - 1\Big| + \Big||y| - 1\Big| \leq 2$$

also forms a polygon.

(a) How many sides has this polygon?

(b) What is its area?

160. Let $\triangle OAB$, $\triangle OCD$, $\triangle OEF$ be equilateral triangles sharing a common vertex O (with a same orientation). Show that the midpoints U, V, W of DE, FA, BC are the vertices of an equilateral triangle; so are the midpoints X, Y, Z of CF, EB, AD.

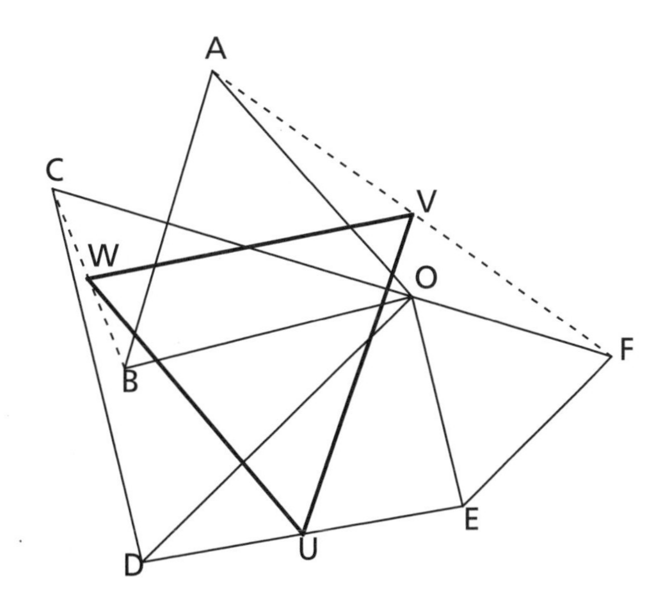

Figure 160

161. Let $x_0 = 1$, $y_0 = 0$, and define x_n and y_n recursively by

$$\begin{cases} x_{n+1} = 2x_n + 3y_n, \\ y_{n+1} = x_n + 2y_n. \end{cases}$$

(a) Show that there exists a coefficient k such that

$$x_{n+1}^2 + ky_{n+1}^2 = x_n^2 + ky_n^2 \quad \text{for all nonnegative integers } n.$$

(b) Show that all the points (x_n, y_n) are on a conic, and find the equation of the conic.

162. For each integer n, let x_n, y_n be a pair of integers satisfying

$$x_n + y_n\sqrt{2} = \left(3 + 2\sqrt{2}\right)^n.$$

Show that all the points (x_n, y_n) are on a conic, and find the equation of the conic.

163. Find the equations of all the straight lines ℓ with the property that whenever a point (x, y) lies on ℓ, then the point $(x + 2y, 4x + 3y)$ also lies on ℓ.

164. Given a point on a line, it is easy to construct the line passing through the given point and perpendicular to the given line. The standard Euclidean construction uses a compass three times. Is this the best possible? In other words, is it possible to construct the desired perpendicular line using a compass fewer than three times?

(There is no restriction on the number of times a straightedge is used.)

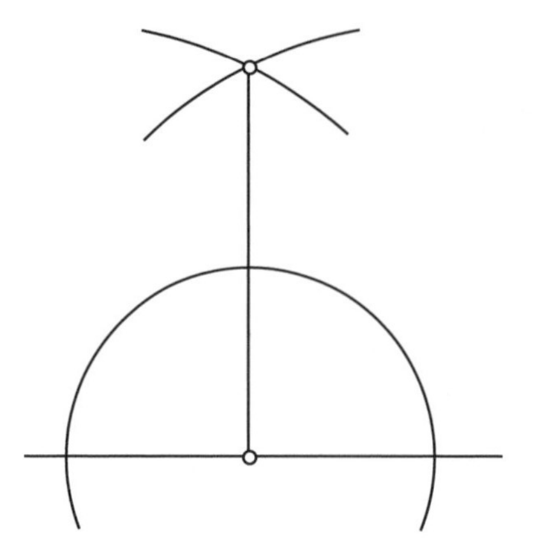

Figure 164

165. Suppose PQ is a line segment on the plane of $\triangle ABC$, whose perpendicular projections on (the extensions of) sides BC, CA, and AB have lengths 2, 3, and 4, respectively. If the lengths of sides BC and CA are 5 and 6, respectively, find the length of the remaining side AB. Is the answer unique?

166. A convex polygon lies inside a closed curve. Show that the perimeter of the convex polygon is not greater than that of the closed curve.

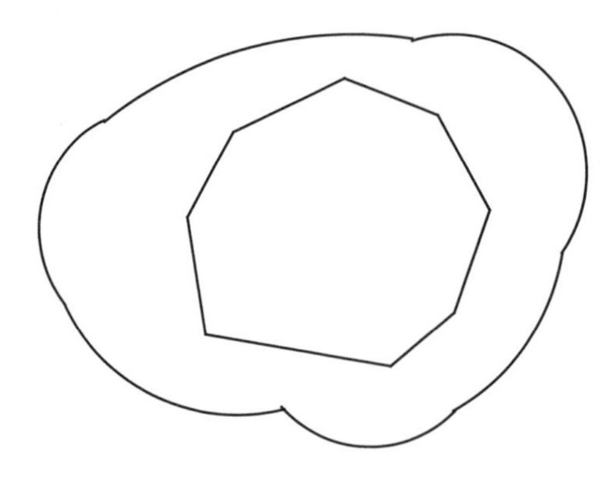

Figure 166

PART TWO Solutions

Chapter 2.1 Number Theory and Algebra

1. The minimum number of "foldings" needed is 29.

   For each folding, the number of pieces increases by 1. Thus, in fact, no matter how you do it, you need 29 foldings.

2. To determine the champion, 49 games have to be played.

   For each game played, one team is eliminated.

3. Darcey got 7 points.

   All together $2 \times \binom{4}{2} = 12$ games must be played. For each game, 2 points are awarded between the two players. Hence Darcey got $2 \times 12 - (6 + 3 + 8) = 7$ points.

4. 24 poles must be removed.

   The least common multiple of 60 $(= 2^2 \cdot 3 \cdot 5)$ and 75 $(= 3 \cdot 5^2)$ is $2^2 \cdot 3 \cdot 5^2 = 300$, and for every 3 meters, we must remove 4 poles.

   Thus, all together we must remove $\frac{18}{3} \times 4 = 24$ poles.

5. Hester has two dimes.

   Let p, n, d and q be the number of pennies, nickels, dimes and quarters, respectively. Then we have

   $$p + n + d + q = 19, \quad p + 5n + 10d + 25q = 93.$$

   Clearly, $p = 3$, 8 or 13.

   (a) Suppose $p = 3$. Then

   $$n + d + q = 16, \quad n + 2d + 5q = 18.$$

   It follows that $d + 4q = 2$. But, by assumption, $d \geq 1$, $q \geq 1$, so this case cannot occur.

(b) Suppose $p = 8$. Then

$$n + d + q = 11, \quad n + 2d + 5q = 17.$$

It follows that $d + 4q = 6$. Because $d \geq 1$, $q \geq 1$, the only possible solution is $d = 2$, $q = 1$ (and $n = 8$).

(c) Suppose $p = 13$. Then

$$n + d + q = 6, \quad n + 2d + 5q = 16.$$

It follows that $d + 4q = 10$. Clearly, $q = 2$ is the only possibility, which gives $d = 2$ (and $n = 2$).

In either case, we have $d = 2$.

6. The merchant sold 23 radios at the price \$86 each.

   Because $1978 = 2 \cdot 23 \cdot 43$, the only factor of 1978 between 149/2 and 149 is $2 \cdot 43 = 86$.

7. $p = 1997$.

   Set $1999p + 1 = n^2$. Then $1999p = (n-1)(n+1)$. Because 1999 and p are primes, both $n - 1$ and $n + 1$ must be primes, and one of them is equal to 1999. Suppose $n - 1 = 1999$; then $p = n + 1 = 2001$, which is not a prime. If $n + 1 = 1999$, then $p = n - 1 = 1997$, which is a prime. Indeed, if $p = 1997$, then

$$1999p + 1 = (1998 + 1)(1998 - 1) + 1 = (1998^2 - 1^2) + 1 = 1998^2.$$

8. (a) 26 people spoke Spanish. (b) 4 people did not speak any of these three languages.

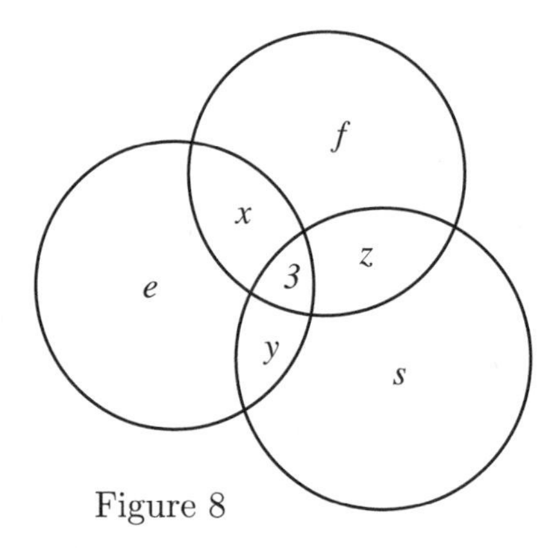

Figure 8

(a) Let e, f, s be the numbers of people who spoke English only, French only, and Spanish only; and x, y, z the numbers of those people who spoke English and French only, English and Spanish only, and French and Spanish only, respectively. Then from the given information we obtain

$$\begin{cases} x + y + z &= 18, \\ e + f + s &= 25, \\ e + x + y + 3 &= 35, \\ f &= 0, \\ x + z + 3 &= 9. \end{cases}$$

Solving the system of simultaneous equations, we obtain

$$x + z = 6, \quad y = 12, \quad e + x = 20.$$

Therefore, the number of people who spoke Spanish is

$$\begin{aligned} s + y + z + 3 &= (e + f + s) + (x + y + z) + 3 - (e + x) \\ &= 25 + 18 + 3 - 20 = 26. \end{aligned}$$

(b) The number of people who spoke at least one of the three languages is

$$(e + f + s) + (x + y + z) + 3 = 25 + 18 + 3 = 46,$$

and so the number of people who did not speak any of these three languages is $50 - 46 = 4$.

9. The magic sum is 175.

   Let s be the magic sum. Adding all seven rows (or all seven columns), we obtain

$$7s = 1 + 2 + 3 + \cdots + 48 + 49 = \frac{49 \times 50}{2}. \quad \therefore\ s = 7 \times 25 = 175.$$

Here is an example for a magic square of order 7.

22	47	16	41	10	35	4
5	23	48	17	42	11	29
30	6	24	49	18	36	12
13	31	7	25	43	19	37
38	14	32	1	26	44	20
21	39	8	33	2	27	45
46	15	40	9	34	3	28

Figure 9

Question: What is the general formula for the magic sum of a magic square of order n (with integers from 1 through n^2)?

10. Label as in the figure on the left.

3	A	B
4	C	D
E	1	F

3	36	2
4	6	9
18	1	12

Figure 10

Then $3 \cdot 4 \cdot E = E \cdot 1 \cdot F$. Because no entry is zero, $E \neq 0$, and so $F = 12$. Similarly,

$$\begin{aligned} 4 \cdot C \cdot D &= 3 \cdot C \cdot F = 3 \cdot C \cdot 12. \quad \therefore\ D = 9. \\ A \cdot C \cdot 1 &= 3 \cdot C \cdot F = 3 \cdot C \cdot 12. \quad \therefore\ A = 36. \end{aligned}$$

Let k be the "magic product"; then

$$\begin{aligned} k &= 3 \cdot A \cdot B = 3 \cdot 36 \cdot B \\ &= 4 \cdot C \cdot D = 4 \cdot C \cdot 9 \\ &= E \cdot 1 \cdot F = E \cdot 1 \cdot 12. \\ \therefore\ k^3 &= (3 \cdot 36 \cdot 4 \cdot 9 \cdot 1 \cdot 12)(BCE) \\ &= (3 \cdot 3^2 \cdot 2^2 \cdot 2^2 \cdot 3^2 \cdot 3 \cdot 2^2)k. \end{aligned}$$

$$\therefore\ k^2 = 2^6 \cdot 3^6, \qquad k = 2^3 \cdot 3^3.$$

Substituting this value of k into the equalities above, we obtain

$$2^3 \cdot 3^3 = 3 \cdot 36 \cdot B = 4 \cdot C \cdot 9 = E \cdot 1 \cdot 12.$$

$$\therefore\ B = 2, \quad C = 2 \cdot 3 = 6, \quad E = 2 \cdot 3^2 = 18.$$

11. Yes, E is the average of all the entries.

Let S be the magic sum, then

$$\begin{aligned} S &= A + E + I = B + E + H = C + E + G. \\ \therefore\ 3S &= (A + B + C) + (G + H + I) + 3E \\ &= 2S + 3E. \\ E &= \frac{1}{3}S = \frac{1}{9}(3S) \\ &= \frac{1}{9}\{(A + B + C) + (D + E + F) + (G + H + I)\}. \end{aligned}$$

A	B	C
D	E	F
G	H	I

Figure 11(c)

Question: Where does this proof fail for magic squares of higher order?

12.

$$\begin{aligned} 1988 &= 281 + \cdots + 284 + \cdots + 287 \\ &= 245 + \cdots + 248 + 249 + \cdots + 252 \\ &= 8 + \cdots + 35 + 36 + \cdots + 63 \end{aligned}$$

To obtain these results, write

$$1988 = a + (a+1) + \cdots + (a+n-1) = na + \frac{n(n-1)}{2};$$
$$\text{i.e.,}\quad n(2a+n-1) = 2 \cdot 1988 = 2^3 \cdot 7 \cdot 71.$$

Because $n < 2a + n - 1$, the only possible factorizations are as follows:

n	$2a+n-1$	a
2	1988	(not an integer)
4	994	(not an integer)
7	568	281
8	497	245
14	284	(not an integer)
28	142	(not an integer)
56	71	8

General Solution. Let us begin by doing an experiment.[1] Can we express 1988 as a sum of two consecutive integers? If so, these two integers must be close to each other (differ by 1); thus, we divide 1988 by 2, and obtain $1988/2 = 994$. Hence

$$\begin{aligned} 994 + 994 &= 1988, \\ \text{but } 993 + 994 &= 1987, \\ 994 + 995 &= 1989. \end{aligned}$$

Therefore, 1988 cannot be expressed as a sum of two consecutive integers.

Can we express 1988 as a sum of three consecutive integers? This time $1988/3$, which is supposed to be the middle term, is not an integer. So the answer is No. How about as a sum of four consecutive integers? Because $1988/4 = 497$, so we try

$$\begin{aligned} 496 + 497 + 498 + 499 &= 1990 \quad \text{(too big)}, \\ 495 + 496 + 497 + 498 &= 1986 \quad \text{(too small)}. \end{aligned}$$

No luck.

How about five terms? Unfortunately, $1988/5$, which should be the middle term, is again not an integer.

From our experiment, we notice that if an integer n is expressible as a sum of consecutive integers with an odd number of terms, say $2k + 1$ terms, then the average (the middle term) m must be an integer, and we obtain

$$\begin{aligned} n &= (2k+1)m \\ &= (m-k) + \cdots + (m-1) + m + (m+1) + \cdots + (m+k). \end{aligned}$$

On the other hand, if an integer is expressible as a sum of consecutive integers with even number of terms, $2m$ terms, say, then the average must lie halfway between

[1] Remember! Even in mathematics, we do need experiments.

two consecutive integers, k and $k+1$, say; i.e., the average is half of an odd integer, $\frac{k+(k+1)}{2} = \frac{2k+1}{2}$. And again we obtain

$$\begin{aligned} n &= \left(\frac{2k+1}{2}\right) \cdot (2m) = (2k+1)m \\ &= (k+1-m) + \cdots + k + (k+1) + \cdots + (k+m). \end{aligned}$$

We conclude that if an integer n is expressible as a sum of two or more consecutive integers, it must have an odd integer factor.

Does this mean that every time we have a factorization of an integer n as $n = (2k+1)m$, we can express n as a sum of consecutive positive integers of

(a) $2k+1$ terms with m as the average; and

(b) $2m$ terms with the average $\frac{k+(k+1)}{2} = \frac{2k+1}{2}$?

The answer turns out to be No. Take a close look at the discussion above. In case (a), we obtain positive integers if and only if $m - k > 0$. On the other hand, for (b), we obtain positive integers if and only if $k + 1 - m > 0$, and these two cases exclude each other. We conclude that for each factorization of n as $n = (2k+1)m$, we have one and only one expression of n as a sum of consecutive positive integers.

Returning to our case $n = 1988 = 2^2 \times 7 \times 71$,

(i) $2k+1 = 7, \quad m = 2^2 \times 71, \quad k = 3 < 284 = m.$

Hence 1988 is expressible as the sum of 7 consecutive positive integers with 284 as the average; i.e.,

$$1988 = 281 + \cdots + 284 + \cdots + 287.$$

(ii) $2k+1 = 71, \quad m = 2^2 \times 7, \quad k = 35 > 28 = m.$

Hence this time we have $2m = 56$ terms with $\frac{2k+1}{2} = \frac{35+36}{2}$ as the average; i.e.,

$$1988 = 8 + \cdots + 35 + 36 + \cdots + 63.$$

(iii) $2k+1 = 7 \times 71, \quad m = 2^2, \quad k = 248 > 4 = m.$

So we obtain an expression containing $2m = 8$ terms with $\frac{2k+1}{2} = \frac{248+249}{2}$ as the average; i.e.,

$$1988 = 245 + \cdots + 248 + 249 + \cdots + 252.$$

Remark. The only positive integers that cannot be expressed as a sum of consecutive positive integers are powers of 2; i.e., integers of the form 2^n.

Exercise. Find the longest string of positive integers whose sum is 3^{10}. What about 3^{11}?

13. The only possible perfect square is (f) 5729581636.

No perfect square can end with 2, 3, 7, 8 (i.e., when divided by 10, a perfect square will not give a remainder of 2, 3, 7 or 8), and so (b), (c), (g), (h) are out.

Every perfect square, when divided by 4, gives remainder 0 or 1, but not 2 or 3 (why?); and to find the remainder (when divided by 4), all we have to do is to test the last two digits (why?); thus (a) and (i) are out.

Every perfect square, when divided by 9, gives remainder 0, 1, 4 or 7 (why?), and to find the remainder (when divided by 9), we can apply the technique of casting out nines. We see that (d) gives the remainder 2, while (j) gives the remainder 6, so these are also out. So only (e) and (f) remain as candidates.

Now for a square of an integer to end with 5, the integer itself must end with 5, and in such a case the last two digits of the square are 25 (why?), so (e) is also out.

The only possible perfect square is (f), and in fact, $5729581636 = 75694^2$.

14. The proposition is correct.

$$\{41, 42, 43, 44, 45, 46, 47, 48, 49, 50\} \text{ and } \{91, 92, 93, 94, 95, 96, 97, 98, 99, 100\}.$$

The last two digits of the square depend on the last two digits of the integer only, so if we can find one set of 10 consecutive integers with the specified property, then we have infinitely many such sets.

Note that the only perfect squares of integers that end with 25 are the ones with 5 as the unit digit. So the last digits of the 10 consecutive integers, if any, must be 1, 2, 3, 4, 5, 6, 7, 8, 9, 0. With a little experiment, we see that the next-to-the-last digit must be either 4 or 9. Thus the two sets of smallest consecutive positive integers with the specified property are

$$\{41, 42, 43, 44, 45, 46, 47, 48, 49, 50\} \text{ and } \{91, 92, 93, 94, 95, 96, 97, 98, 99, 100\}.$$

15. No, there exists no such perfect square.

It is easy to verify the following propositions:

(a) No perfect square ends with 3 or 7.

(b) If a perfect square ends with 5, then it must end with 25.

Hence in our case, the only possible perfect square must end with 1 or 9. That is, the last two digits of our possible perfect square must be 11, 31, 51, 71, 91, 19, 39, 59, 79, or 99. But all such integers give remainder 3 when divided by 4. However,

(c) Every perfect square, when divided by 4, gives remainder 0 or 1 (not 2 or 3).

Hence no such perfect square exists.

Remark. Our proof shows that for an integer to be a perfect square, at least one of the last two digits must be even.

Alternate Solution. Clearly, if an integer satisfying the conditions is a perfect square, then it must be a perfect square of an odd integer, $10m+n$, say, where m is a positive integer and n is 1, 3, 5, 7, or 9. (For example, to represent 1991, we set $m = 199$, $n = 1$.) Now,

$$(10m + n)^2 = 100m^2 + 20mn + n^2.$$

As a multiple of 100, $100m^2$ has no effect on the last two digits, and $20mn$ is an even multiple of 10, so the parity (even or odd) of the digit next to the last is determined by that of n^2. Because $n = 1, 3, 5, 7$, or 9, by checking

$$1^2 = 01, \quad 3^2 = 09, \quad 5^2 = 25, \quad 7^2 = 49, \quad 9^2 = 81,$$

we see that if the last digit of a perfect square is odd, then the digit next to the last must be even.

16. Both of their assertions are correct.

(a) Suppose the last digit of n^2 is 6; then the integer n itself must end with 4 or 6; i.e., $n = 10k + 4$ or $n = 10k + 6$ for some nonnegative integer k.

$$\begin{aligned} \therefore\ n^2 &= 100k^2 + 80k + 16 = (10k^2 + 8k + 1) \times 10 + 6, \\ \text{or}\quad n^2 &= 100k^2 + 120k + 36 = (10k^2 + 12k + 3) \times 10 + 6. \end{aligned}$$

Because both $10k^2 + 8k + 1$ and $10k^2 + 12k + 3$ are odd for any integer k, Veronica's assertion is correct.

Alternate Solution. If a perfect square n^2 ends with 6, then n itself must be even; i.e., $n = 2m$ for some integer m. But then $n^2 = 4m^2$ is divisible by 4. Now, if the next-to-the-last digit of n^2 is even (and the last digit is 6), then n^2 is not divisible by 4. Thus, the next-to-the-last digit of n^2 must be odd if the last digit is 6.

(b) A perfect square must end with 0, 1, 4, 5, 6 or 9. Now, if the next-to-the-last digit is odd, and the last digit is 0, 1, 4, 5 or 9, then the remainder when divided by 4, must be 2 or 3. But a perfect square, when divided by 4, must give a remainder 0 or 1. ($\because\ (2n)^2 = 4n^2, \quad (2n-1)^2 = 4(n^2 - n) + 1$.) So, when the next-to-the-last digit is odd, the only possibility for the last digit is 6; and, in fact, there are infinitely many such perfect squares. For example, $16 = 4^2$, $36 = 6^2$, $196 = 14^2$, $256 = 16^2, \cdots$.

Thus, Paige's assertion is also correct.

17. (a) The remainder is 7. (b) $A = B = 1, \quad n = 4$.

(a) By the method of casting out nines, we see that 1993 gives the remainder 4 when divided by 9; i.e.,

$$\begin{aligned} 1993 &= 9k + 4 \quad \text{for some integer } k. \\ \therefore\ 1993^2 &= (9k+4)^2 = 9^2k^2 + 2 \cdot 9 \cdot 4k + 4^2 \\ &= 9m + 7 \quad (m = 9k^2 + 8k + 1). \end{aligned}$$

Therefore, the desired remainder is 7.

Remark. If we use the notation $a \equiv b \pmod{m}$ to denote that a and b give the same remainder when divided by m (in other words, $a \equiv b \pmod{m}$ means that $a - b$ is a multiple of m), then the solution above can be rewritten as

$$1993 \equiv 4 \pmod{9}, \quad \therefore\ 1993^2 \equiv 4^2 \equiv 16 \equiv 7 \pmod{9}.$$

(b) Note that

$$10^{13} < 15777A7325840B < 1.6 \times 10^{13},$$

and

$$10^3 < 1993 < 2 \times 10^3, \quad \therefore\ 10^{12} < 1993^4 < 1.6 \times 10^{13}.$$

It follows that $n = 4$ ($\because\ 1993^3 < 8 \times 10^9 < 10^{12}$, and $1993^5 > 10^{15} > 1.6 \times 10^{13}$).

Now, the last digit B must be the last digit of 3^4; i.e.,

$$B \equiv 15777A7325840B \equiv 1993^4 \equiv 3^4 \equiv 81 \equiv 1 \pmod{10}.$$

But $0 \le B \le 9$, $\therefore\ B = 1$.

Similarly, to determine A,

$$15777A73258401 \equiv 3 + A \pmod 9,$$

and

$$1993^4 \equiv 4^4 \equiv 16^2 \equiv 7^2 \equiv 4 \pmod 9.$$

$$\therefore\ 3 + A \equiv 4 \pmod 9\text{; i.e., } A \equiv 1 \pmod 9.$$

But $0 \le A \le 9$, $\therefore\ A = 1$.

18. $n = 87$.

Because $10^9 < n^5 < 10^{10}$, n must be a 2-digit number; and by a little experiment we know that if $n = 10m + k$ for some nonnegative integers m and k, where $0 \le k \le 9$ (i.e., k is the unit digit of n), then $n^5 \equiv k \pmod{10}$. Therefore the last digit (unit digit) of n must be 7. Moreover, by the method of casting out nines,

$$4984209207 \equiv 0 \pmod 9,$$

and so we have $n \equiv 0 \pmod 3$. It follows that $n = 27$, 57, or 87. However,

$$\begin{aligned} 57 &< 60, \\ 57^2 &< 60^2 = 36 \times 10^2 < 4 \times 10^3, \\ 57^4 &< 60^4 < 16 \times 10^6 < 2 \times 10^7, \\ \therefore\ 57^5 &< 60 \times 2 \times 10^7 = 12 \times 10^8 < 4.9 \times 10^9 < n^5. \end{aligned}$$

Hence $n > 60$. $\therefore\ n = 87$.

19. Suppose the number chosen is

$$a \cdot 10^2 + b \cdot 10 + c \quad (a \ne c).$$

Without loss of generality, we may assume $a > c$. Then the (absolute value of) the difference of this number and the number obtained by reversing the order of the digits is

$$\begin{aligned} &(a \cdot 10^2 + b \cdot 10 + c) - (c \cdot 10^2 + b \cdot 10 + a) \\ &\quad = (a - 1 - c) \cdot 10^2 + (9 \cdot 10) + (10 + c - a). \end{aligned}$$

If we add this number and the number obtained by reversing the order of the digits, we obtain

$$\begin{aligned} &\{(a-1-c)\cdot 10^2+(9\cdot 10)+(10+c-a)\} \\ &\quad +\{(10+c-a)\cdot 10^2+(9\cdot 10)+(a-1-c)\} \\ =\ &(10-1)\cdot 10^2+2\cdot(9\cdot 10)+(10-1)=1089, \end{aligned}$$

which is independent of the choice of the original number.

20. Answer: 1089.

Let a and b be the first two numbers. Then the first 10 numbers in the sequence are:

$$a,\ b,\ a+b,\ a+2b,\ 2a+3b,\ 3a+5b,\ 5a+8b,\ 8a+13b,\ 13a+21b,\ 21a+34b,$$

and their sum is $55a+88b$, which is 11 times the seventh number.

21. (a) 5. (b) The difference obtained is always divisible by 9.

Again, let us do some experiments. Start from a small number, say 95. Rearrange its digits to obtain 59. Then $95-59=36$. So, if we say 6, then the answer should be 3, while if we say 3, then the answer should be 6. We don't quite see a pattern, so let's try a 3-digit number, say 123. Scramble its digits to obtain 231. Then $231-123=108$. If we say 1, then the answer should be 8, while if we say 8, then the answer should be 1. Now, it becomes suspicious that these digits should add up to 9. But what about for larger numbers? Let's take 1995. Scramble its digits to obtain 9951, say. Then $9951-1995=7956$. This time the digits add up to $7+9+5+6=27$, a multiple of 9. This means, by the method of casting out nines, 7956 itself must be a multiple of 9. And in our previous examples, 36 and 108 are all multiples of 9 too.

This is trivial, because for any number, by the method of casting out nines, the remainders when divided by 9, for the sum of the digits and for the original number are the same. Therefore, no matter how you scramble its digits, the original number and a number obtained by scrambling have the same remainder when divided by 9. It follows that their difference is always divisible by 9, and all we have to do is supply a digit so that the sum of the digits is a multiple of 9. This explains why the digits in the difference can be presented in any order, and 0 is excluded. Now the rest is trivial and we let the reader finish the solution.

22. (a) 680. (b) $\{715, 364, 924\}$ or $\{-286, 364, -77\}$.

Note that 70 is a multiple of 5 and 7, but gives a remainder 1 when divided by 3. Similarly, 21 is a multiple of 3 and 7, but gives a remainder 1 when divided by 5; and 15 is a multiple of 3 and 5, but gives a remainder 1 when divided by 7. That is why $70\times 2+21\times 4+15\times 5$ ($=299$) gives the remainder 2, 4 and 5 when divided by 3, 5 and 7, respectively.

So, the first number in Nicky's magic triple must be a multiple of 11 and 13, which would have a remainder 1 when divided by 7. Because

$$11\times 13\equiv 4\times(-1)\equiv -4\equiv 3 \pmod 7,$$

we obtain

$$286 \equiv 2 \times (11 \times 13) \equiv 2 \times 3 \equiv 6 \equiv -1 \pmod 7, \quad \therefore -286 \equiv 1 \pmod 7.$$

We could keep -286 as the first number in Nicky's magic triple, but in case we prefer a positive number, then we can have

$$-286 + 7 \times 11 \times 13 = -286 + 1001 = 715.$$

Similarly, the second number in Nicky's magic triple must be a multiple of 7×13, which would have a remainder 1 when divided by 11. Because

$$7 \times 13 \equiv 7 \times 2 \equiv 14 \equiv 3 \pmod{11},$$

we have

$$364 \equiv 4 \times (7 \times 13) \equiv 4 \times 3 \equiv 12 \equiv 1 \pmod{11};$$

the second number in Nicky's magic triple is 364.

Finally, the third number in Nicky's magic triple must be a multiple of 7×11, which would have a remainder 1 when divided by 13. Because

$$7 \times 11 \equiv 7 \times (-2) \equiv -14 \equiv -1 \pmod{13},$$

we obtain

$$-77 \equiv 1 \pmod{13}.$$

Again we could keep -77 as the third number in Nicky's magic triple, or choose

$$-77 + 7 \times 11 \times 13 = -77 + 1001 = 924.$$

Thus Nicky's magic triple is $\{715, 364, 924\}$, where 715 may be replaced by -286, and similarly, 924 by -77.

Therefore, Uncle Paul's secret number is

$$\begin{aligned} 715 \times 1 + 364 \times 9 + 924 \times 4 &= 715 + 3276 + 3696 \\ &= 7687 \equiv 680 \pmod{1001}; \end{aligned}$$

equivalently,

$$\begin{aligned} (-286) \times 1 + 364 \times 9 + (-77) \times 4 &= -286 + 3276 - 308 \\ &= 2682 \equiv 680 \pmod{1001}. \end{aligned}$$

Exercise. What would the 'magic quadruple' be if the divisors are 5, 7, 11, and 17?

23. $\{158, 159, 160, 161\}$.

Let the four consecutive positive integers be $5n-2$, $5n-1$, $5n$, $5n+1$. Because $5n-2$ must be divisible by 2, n must be an even number. Set $n = 2k$. Then the four consecutive positive integers are $10k-2$, $10k-1$, $10k$, $10k+1$. Now, $10k-1$ must be divisible by 3, so k must be of the form $k = 3t+1$; and the four consecutive positive integers are $30t+8$, $30t+9$, $30t+10$, $30t+11$. Finally, because $30t+11$ is a multiple of 7, t must be of the form $t = 7m+5$, and the four consecutive positive integers are

$$210m+158,\ 210m+159,\ 210m+160,\ 210m+161.$$

Substituting $m = 0$, we obtain the smallest consecutive positive integers 158, 159, 160, 161.

Alternate Solution. Let four consecutive integers be $7k-3$, $7k-2$, $7k-1$, $7k$ $(k \geq 1)$. Then

$$\begin{array}{lllll} 7k-3 \equiv 0 & \pmod 2; & \text{i.e.,} & k \equiv 1 & \pmod 2, \\ 7k-2 \equiv 0 & \pmod 3; & \text{i.e.,} & k \equiv 2 & \pmod 3, \\ 7k-1 \equiv 0 & \pmod 5; & \text{i.e.,} & 2k \equiv 1 & \pmod 5. \end{array}$$

Multiplying the last congruence by 3, we obtain

$$k \equiv 6k \equiv 3 \pmod 5; \text{i.e.,}\ k = 5m+3 \quad (\text{for some positive integer } m).$$

Substituting this result into the second congruence above, we obtain

$$5m+3 \equiv 2; \text{i.e.,}\ m \equiv 1 \pmod 3.$$

$$\begin{aligned} \therefore m &= 3n+1 \text{ (for some } n), \\ k &= 5(3n+1)+3 = 15n+8 \equiv 1 \pmod 2. \\ \therefore n &= 2t+1 \text{ (for some } t), \\ k &= 15(2t+1)+8 = 30t+23. \end{aligned}$$

Hence four consecutive integers are

$$7(30t+23)-3,\ 7(30t+23)-2,\ 7(30t+23)-1,\ 7(30t+23);$$

i.e., the general solution is

$$210t+158,\ 210t+159,\ 210t+160,\ 210t+161.$$

Set $t = 0$ to obtain the smallest integers with the required properties.

24. $x = 1,\ y = 1,\ z = 3.$

Because x, y, z are positive integers,

$$\begin{aligned} 1 &\leq x < x + \frac{y}{19} + \frac{z}{97} = \frac{1997}{19 \times 97} < 2, \quad \therefore x = 1. \\ \frac{1}{19} &\leq \frac{y}{19} < \frac{y}{19} + \frac{z}{97} = \frac{1997}{19 \times 97} - 1 = \frac{154}{1843} < \frac{2}{19}, \quad \therefore y = 1. \\ \frac{z}{97} &= \frac{154}{19 \times 97} - \frac{1}{19} = \frac{154-97}{19 \times 97} = \frac{57}{19 \times 97} = \frac{3}{97}, \quad \therefore z = 3. \end{aligned}$$

Exercise. Find integers a, b_1, b_2, b_3, c such that

$$\frac{a}{2} + \frac{b_1}{3} + \frac{b_2}{3^2} + \frac{b_3}{3^3} + \frac{c}{37} = \frac{1997}{1998},$$

where $0 \leq a < 2$, $0 \leq b_j < 3$ $(j = 1, 2, 3)$, $0 \leq c < 37$.

25. (a) $p = 5$, $q = 27$. (b) $r = 11$, $s = 74$.

Let $x = 0.\dot{1}8\dot{5} = 0.185185185\cdots$; then $1000x = 185.185185\cdots$. The difference of these two equalities gives

$$999x = 185; \quad \therefore\ x = \frac{185}{999} = \frac{5 \cdot 37}{3^3 \cdot 37} = \frac{5}{27}.$$

Because 5 and 27 are relatively prime, we conclude that $p = 5$, $q = 27$.

(b) The same method works for this case too. Let $y = 0.1\dot{4}8\dot{6} = 0.1486486486\cdots$; then $1000y = 148.648648\cdots$. Their difference gives

$$999y = 148.5; \quad \therefore\ y = \frac{148.5}{999} = \frac{1485}{9990} = \frac{5 \cdot 3^3 \cdot 11}{2 \cdot 5 \cdot 3^3 \cdot 37} = \frac{11}{74}.$$

Because 11 and 74 are relatively prime, we conclude that $r = 11$, $s = 74$.

26. (a) 27, 37, 111, 333 and 999. (b) $37\,\overline{)\,6\,}$.

(a) Suppose $x = n.\dot{a}b\dot{c}$. Then $1000x = m.\dot{a}b\dot{c}$ ($m = 10^3 n + 10^2 a + 10b + c$), and the difference of these two equalities gives

$$999x = k, \quad x = \frac{k}{999} \quad \text{(where } k = 999n + 10^2 a + 10b + c \text{ is an integer)}.$$

So, if $\frac{p}{q} = \frac{k}{999}$ (where p and q have no common divisor other than 1), then q must be a divisor of 999. On the other hand, if q is a divisor of 999, but not of 99, then q, as the denominator, gives repeating decimal of minimum length 3. (What if q is a divisor of 99?) Because $999 = 3^3 \cdot 37$, the only possibilities for q are 27, 37, 111, 333 and 999.

(b) Let us label as in Figure 26(a). Because the division produces repeating decimal, we must have $C = G$. From our discussion in Part (a), we know $AB = 27$ or $AB = 37$.

(i) Suppose $AB = 27$. Then EF must be a multiple of 27, and so $EF = 27$, 54 or 81. Clearly, $D = 0$, and so if $EF = 27$, then $C = G = 3$; if $EF = 54$, then $C = G = 6$; if $EF = 81$, then $C = G = 9$. Thus these three cases result in

$$\frac{C}{AB} = \frac{3}{27} = \frac{1}{9}, \quad \frac{C}{AB} = \frac{6}{27} = \frac{2}{9}, \quad \frac{C}{AB} = \frac{9}{27} = \frac{1}{3};$$

and in all these cases, we obtain repeating decimals of minimum length 1 (not 3). So the case $AB = 27$ can not occur.

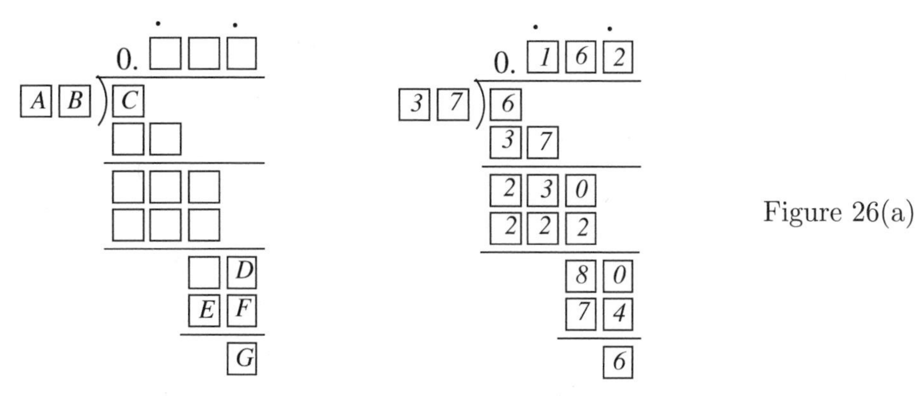

Figure 26(a)

(ii) Suppose $AB = 37$. Then as before, EF is a multiple of 37, and hence must be either $EF = 37$ or 74. Suppose $EF = 37$; then because $D = 0$, we must have $C = G = 3$, and the division becomes $37\overline{)\ 3\ }$. A simple trial shows that this does not fit the given pattern. Hence $EF = 74$ is the only remaining case. This time (because $D = 0$), we have $C = G = 6$. A simple trial shows that this fits the pattern perfectly.

27. 34 votes will guarantee the student being elected.

All together, there are $2 \times 100 = 200$ votes. Let x be the minimum number of votes that guarantees the student being elected. Then we have

$$x \geq \frac{200 - x}{5},$$

which gives $x \geq 34$.

28. The minimum value is $\sqrt{\frac{24}{6}} + \sqrt{\frac{96}{24}} = 4$.

The numbers in the sequence can be expressed as

$$\sqrt{\frac{x}{6}} + \sqrt{\frac{96}{x}} \quad (x = 7, 8, \cdots, 95).$$

Now the inequality between the arithmetic mean and the geometric mean (see Problem 152) says

$$a + b \geq 2\sqrt{ab} = 2(ab)^{1/2} \quad (\text{whenever } a \geq 0,\ b \geq 0),$$

and equality holds if and only if $a = b$.

Substituting $a = \sqrt{x/6}$, $b = \sqrt{96/x}$, we obtain

$$\begin{aligned} \sqrt{\frac{x}{6}} + \sqrt{\frac{96}{x}} &\geq 2\left(\sqrt{\frac{x}{6}} \cdot \sqrt{\frac{96}{x}}\right)^{1/2} = 2\left(\sqrt{\frac{96}{6}}\right)^{1/2} \\ &= 2(\sqrt{16})^{1/2} = 2(4)^{1/2} = 2 \cdot 2 = 4. \end{aligned}$$

Equality holds if and only if

$$\sqrt{\frac{x}{6}} = \sqrt{\frac{96}{x}}; \text{ i.e., } x = \sqrt{6 \cdot 96} = 6 \cdot 4 = 24.$$

Hence $\sqrt{\frac{24}{6}} + \sqrt{\frac{96}{24}} = 4$ is the minimum value.

Remark. Solving this problem using calculus is like killing a fly with a sledge-hammer.

29. $3 < x < 4$.

$$\begin{aligned}
\frac{x-1}{x-3} \geq \frac{x-2}{x-4} &\iff \frac{x-1}{x-3} - \frac{x-2}{x-4} \geq 0 \\
&\iff \frac{(x-1)(x-4)-(x-2)(x-3)}{(x-3)(x-4)} \geq 0 \\
&\iff \frac{-2}{(x-3)(x-4)} \geq 0 \\
&\iff (x-3)(x-4) < 0 \quad (\text{why not "}\leq\text{" ?}) \\
&\iff 3 < x < 4.
\end{aligned}$$

•

Question: What's wrong with the following "reasoning"?

Cross multiplying the given inequalities to get rid of the denominators, we obtain

$$(x-1)(x-4) \geq (x-2)(x-3);$$

i.e.,

$$x^2 - 5x + 4 \geq x^2 - 5x + 6. \quad \therefore\ 4 \geq 6.$$

But this is absurd, so there exists no real number x which satisfies the given inequality.

30. $n = 20$.

Because

$$\frac{1}{\sqrt{k}+\sqrt{k+1}} < \frac{1}{2\sqrt{k}} < \frac{1}{\sqrt{k-1}+\sqrt{k}},$$

we have

$$2\left(\sqrt{k+1}-\sqrt{k}\right) < \frac{1}{\sqrt{k}} < 2\left(\sqrt{k}-\sqrt{k-1}\right).$$

From the inequality on the left,

$$\sum_{k=1}^{120} \frac{1}{\sqrt{k}} > 2\sum_{k=1}^{120}\left(\sqrt{k+1}-\sqrt{k}\right) = 2\left(\sqrt{121}-\sqrt{1}\right) = 20.$$

From the inequality on the right,

$$\begin{aligned}
\sum_{k=1}^{120} \frac{1}{\sqrt{k}} &= \frac{1}{\sqrt{1}} + \sum_{k=2}^{120} \frac{1}{\sqrt{k}} < 1 + 2\sum_{k=2}^{120}\left(\sqrt{k}-\sqrt{k-1}\right) \\
&= 1 + 2\left(\sqrt{120}-\sqrt{1}\right) = 2\sqrt{120} - 1 \\
&< 2\sqrt{121} - 1 = 2 \cdot 11 - 1 = 21. \\
\therefore\ 20 &< \sum_{k=1}^{120} \frac{1}{\sqrt{k}} < 21.
\end{aligned}$$

Alternate Solution (with calculus).

Because $f(x) = \dfrac{1}{\sqrt{x}}$ is a monotone decreasing function for $x > 0$, we have

$$\int_k^{k+1} \frac{dx}{\sqrt{x}} < \frac{1}{\sqrt{k}} < \int_{k-1}^{k} \frac{dx}{\sqrt{x}} \quad (k \geq 1).$$

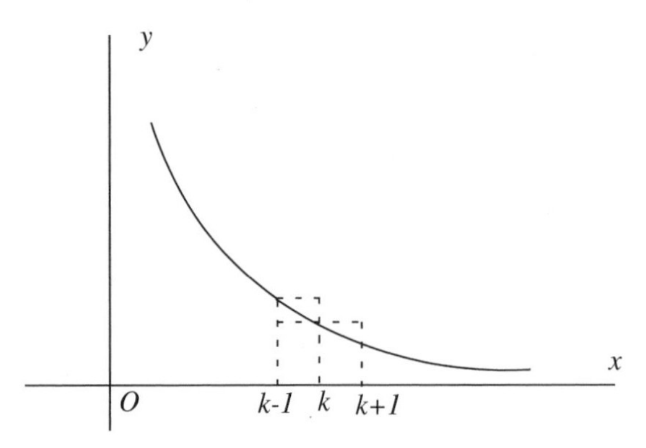

Figure 30

From the inequality on the left, we have

$$\sum_{k=1}^{120} \frac{1}{\sqrt{k}} > \sum_{k=1}^{120} \int_k^{k+1} \frac{dx}{\sqrt{x}} = \int_1^{121} \frac{dx}{\sqrt{x}} = 2\left[\sqrt{x}\right]_1^{121} = 2\left(\sqrt{121} - \sqrt{1}\right) = 20.$$

From the inequality on the right, we have

$$\begin{aligned} \sum_{k=1}^{120} \frac{1}{\sqrt{k}} &= 1 + \sum_{k=2}^{120} \frac{1}{\sqrt{k}} < 1 + \sum_{k=2}^{120} \int_{k-1}^{k} \frac{dx}{\sqrt{x}} \\ &= 1 + \int_1^{120} \frac{dx}{\sqrt{x}} < 1 + \int_1^{121} \frac{dx}{\sqrt{x}} = 1 + 20 = 21. \\ \therefore\ 20 &< \sum_{k=1}^{120} \frac{1}{\sqrt{k}} < 21. \end{aligned}$$

31. $k = 97$.

Because

$$0 < \frac{1}{\sqrt{3} - \sqrt{2}} = \sqrt{3} + \sqrt{2},$$

the given inequalities are equivalent to

$$k < \left(\sqrt{3} + \sqrt{2}\right)^4 < k + 1.$$

$$\begin{aligned} \left(\sqrt{3} + \sqrt{2}\right)^2 &= 5 + 2\sqrt{6}, & \left(\sqrt{3} - \sqrt{2}\right)^2 &= 5 - 2\sqrt{6}, \\ \left(\sqrt{3} + \sqrt{2}\right)^4 &= 49 + 20\sqrt{6}, & \left(\sqrt{3} - \sqrt{2}\right)^4 &= 49 - 20\sqrt{6}. \end{aligned}$$

$$\therefore \left(\sqrt{3} + \sqrt{2}\right)^4 + \left(\sqrt{3} - \sqrt{2}\right)^4 = 2 \cdot 49 = 98.$$

But $0 < (\sqrt{3}-\sqrt{2})^4 < 1$, and so

$$97 < \left(\sqrt{3}+\sqrt{2}\right)^4 < 98; \text{ i.e., } \quad \frac{1}{98} < \left(\sqrt{3}-\sqrt{2}\right)^4 < \frac{1}{97}.$$

Exercise. Suppose x_n and y_n are the integer and the fractional parts of $\sqrt{3}+\sqrt{2}$, respectively; i.e., x_n is the integer satisfying

$$x_n < \left(\sqrt{3}+\sqrt{2}\right)^n < x_n + 1 \quad \text{and} \quad 0 < y_n < 1.$$

Prove or disprove: $0 < y_n < \frac{1}{2}$ if n is odd; and $\frac{1}{2} < y_n < 1$ if n is even.

32. (a) $a = 1, b = 2$. (d) $k_5 = 82$.

(a) In Figure 32 we have $\angle ABC = \frac{\pi}{2}$, $\overline{AB} = \overline{BC}$, $\overline{AC} = \overline{CD}$.

$$\begin{aligned} \therefore\ \angle BAD &= \angle BAC + \angle CAD = \frac{\pi}{4} + \frac{\pi}{8} = \frac{3\pi}{8}. \\ \tan\frac{3\pi}{8} &= \frac{\overline{BD}}{\overline{AB}} = 1+\sqrt{2}. \quad \therefore\ a = 1,\ b = 2. \end{aligned}$$

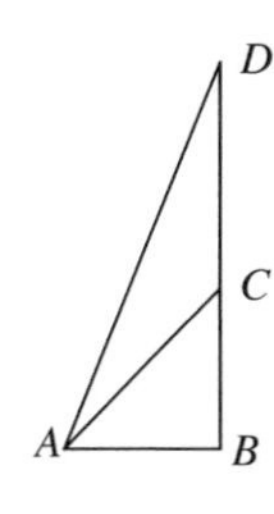

Figure 32

Alternatively,

$$\begin{aligned} \tan\frac{3\pi}{8} &= \tan\left(\frac{\pi}{4}+\frac{\pi}{8}\right) = \frac{\tan\frac{\pi}{4} + \tan\frac{\pi}{8}}{1 - \tan\frac{\pi}{4}\cdot\tan\frac{\pi}{8}} \\ &= \frac{1 + (\sqrt{2}-1)}{1 - (\sqrt{2}-1)} \quad \text{(by Problem 158)} \\ &= \frac{\sqrt{2}}{2-\sqrt{2}} = \frac{1}{\sqrt{2}-1} = \sqrt{2}+1. \end{aligned}$$

(b) $\cot\frac{3\pi}{8} = \frac{1}{1+\sqrt{2}} = \sqrt{2}-1.$

$$\therefore\ \left(\tan\frac{3\pi}{8}\right)^n + (-1)^n\left(\cot\frac{3\pi}{8}\right)^n = (1+\sqrt{2})^n + (1-\sqrt{2})^n.$$

Let p_n and q_n be (nonnegative) integers satisfying

$$(1+\sqrt{2})^n = p_n + q_n\sqrt{2}.$$

Then

$$(1-\sqrt{2})^n = p_n - q_n\sqrt{2}.$$

$$\therefore \left(\tan\frac{3\pi}{8}\right)^n + (-1)^n\left(\cot\frac{3\pi}{8}\right)^n = 2p_n$$

is an even integer.

(c) Clearly, $0 < \cot\dfrac{3\pi}{8} = \sqrt{2}-1 < 1$. Thus, if n is an even integer, then from Part (b),

$$2p_n - 1 < \left(\tan\frac{3\pi}{8}\right)^n < 2p_n.$$

Therefore, $k_n = 2p_n - 1$ is an odd integer. If n is an odd integer, then

$$2p_n < \left(\tan\frac{3\pi}{8}\right)^n < 2p_n + 1.$$

Therefore, $k_n = 2p_n$ is an even integer.

(d) We have

$$\begin{aligned}
p_1 + q_1\sqrt{2} &= 1+\sqrt{2}, \quad \therefore\ p_1 = q_1 = 1.\\
p_2 + q_2\sqrt{2} &= (1+\sqrt{2})^2 = 3+2\sqrt{2}, \quad \therefore\ p_2 = 3,\ q_2 = 2.\\
p_4 + q_4\sqrt{2} &= (1+\sqrt{2})^4 = (3+2\sqrt{2})^2\\
&= 17+12\sqrt{2}, \quad \therefore\ p_4 = 17,\ q_4 = 12.\\
p_5 + q_5\sqrt{2} &= (1+\sqrt{2})^5 = (17+12\sqrt{2})(1+\sqrt{2})\\
&= (17+24)+(17+12)\sqrt{2}\\
&= 41+29\sqrt{2}, \quad \therefore\ p_5 = 41,\ q_5 = 29.
\end{aligned}$$

From Part (c), $k_5 = 2p_5 = 2\cdot 41 = 82$.

Exercise. Show that, for every nonnegative integer n,

(a) $p_{n+1} = p_n + 2q_n,\ q_{n+1} = p_n + q_n$.

(b)

$$\begin{aligned}
p_n &= \frac{1}{2}\left\{(1+\sqrt{2})^n + (1-\sqrt{2})^n\right\},\\
q_n &= \frac{1}{2\sqrt{2}}\left\{(1+\sqrt{2})^n - (1-\sqrt{2})^n\right\}.
\end{aligned}$$

Alternate Solution. Let $x = \tan\dfrac{3\pi}{8} = \sqrt{2}+1$. Then $\dfrac{1}{x} = \cot\dfrac{3\pi}{8} = \sqrt{2}-1$.

$$\begin{aligned}
\therefore\ \tan\frac{3\pi}{8} - \cot\frac{3\pi}{8} &= x - \frac{1}{x} = 2.\\
\left(\tan\frac{3\pi}{8}\right)^2 + \left(\cot\frac{3\pi}{8}\right)^2 &= x^2 + \frac{1}{x^2}\\
&= \left(x-\frac{1}{x}\right)^2 + 2 = 2^2 + 2 = 6.
\end{aligned}$$

$$\begin{aligned}
\left(\tan\frac{3\pi}{8}\right)^3-\left(\cot\frac{3\pi}{8}\right)^3 &= x^3-\frac{1}{x^3}\\
&= \left(x-\frac{1}{x}\right)\left(x^2+\frac{1}{x^2}\right)+\left(x-\frac{1}{x}\right)\\
&= 2(6+1)=14.\\
\left(\tan\frac{3\pi}{8}\right)^5-\left(\cot\frac{3\pi}{8}\right)^5 &= x^5-\frac{1}{x^5}\\
&= \left(x^3-\frac{1}{x^3}\right)\left(x^2+\frac{1}{x^2}\right)-\left(x-\frac{1}{x}\right)\\
&= 14\cdot 6-2=82.
\end{aligned}$$

(See Problem 51.)

$$\therefore\ 82<\left(\tan\frac{3\pi}{8}\right)^5<83\quad \left(\because\ 0<\frac{1}{x}=\cot\frac{3\pi}{8}=\sqrt{2}-1<1\right).$$

33. $p=5,\ q=4,\ r=6,\ s=5.$

Again let us experiment. From the magic formula, the next K is given by

$$K=11\cdot 9^2+20\cdot 9\cdot 11+9\cdot 11^2=3960,$$

and

$$2K+1=7921=89^2,\quad 3K+1=11881=109^2.$$

It is not hard to confirm that

$$89=5\cdot 9+4\cdot 11,\quad 109=6\cdot 9+5\cdot 11.$$

Thus we suspect that $p=5$, $q=4$, $r=6$, $s=5$. Note also that when $k=0$, we have $2k+1=1^2$, $3k+1=1^2$ ($\therefore\ m=n=1$), and the magic formula gives

$$\begin{aligned}
K &= 11\cdot 1^2+20\cdot 1\cdot 1+9\cdot 1^2=40,\\
2K+1 &= 9^2=(5\cdot 1+4\cdot 1)^2,\\
3K+1 &= 11^2=(6\cdot 1+5\cdot 1)^2.
\end{aligned}$$

To prove that our conjecture is always true; i.e.,

$$\begin{aligned}
2K+1 &= 2\left(11m^2+20mn+9n^2\right)+1=(5m+4n)^2,\\
3K+1 &= 3\left(11m^2+20mn+9n^2\right)+1=(6m+5n)^2,
\end{aligned}$$

observe that the right members of these two equalities are quadratic homogeneous expressions of m and n. So we must express 1 in terms of m and n. But

$$\begin{aligned}
1 &= 3(2k+1)-2(3k+1)=3m^2-2n^2.\\
\therefore\ 2K+1 &= 2\left(11m^2+20mn+9n^2\right)+1\\
&= 2\left(11m^2+20mn+9n^2\right)+(3m^2-2n^2)\\
&= 25m^2+40mn+16n^2=(5m+4n)^2,\\
3K+1 &= 3\left(11m^2+20mn+9n^2\right)+1\\
&= 3\left(11m^2+20mn+9n^2\right)+(3m^2-2n^2)\\
&= 36m^2+60mn+25n^2=(6m+5n)^2.
\end{aligned}$$

Remark. In fact, all the integers k with the specified property are (with obvious notations) given by

$$k_j = \frac{1}{24}\left\{\left(\sqrt{3}+\sqrt{2}\right)^{4j+2} + \left(\sqrt{3}-\sqrt{2}\right)^{4j+2} - 10\right\} \quad (j = 0, 1, 2, \cdots).$$

Moreover,

$$\begin{aligned} m_j &= \frac{1}{2\sqrt{3}}\left\{\left(\sqrt{3}+\sqrt{2}\right)^{2j+1} + \left(\sqrt{3}-\sqrt{2}\right)^{2j+1}\right\}, \\ n_j &= \frac{1}{2\sqrt{2}}\left\{\left(\sqrt{3}+\sqrt{2}\right)^{2j+1} - \left(\sqrt{3}-\sqrt{2}\right)^{2j+1}\right\}. \end{aligned}$$

j	0	1	2	3	$\cdots$
k	0	40	3960	388080	$\cdots$
m	1	9	89	881	$\cdots$
n	1	11	109	1079	$\cdots$

Exercise. What if we replace $2k+1$ and $3k+1$ by, say, $3k+1$ and $8k+1$?
Hint: Try $K = \frac{1}{5}\left(39m^2 + 50mn + 16n^2\right)$.
Question. Can you figure out how the magic formulas were found?

34. $k = \frac{1}{2}, -1.$

$$k = \frac{a}{b+c} = \frac{b}{c+a} = \frac{c}{a+b}.$$

$$\therefore a = k(b+c), \quad b = k(c+a), \quad c = k(a+b).$$

Adding the last three equalities, we obtain

$$a+b+c = 2k(a+b+c).$$

Therefore, if $a+b+c \neq 0$, then dividing both sides[2] by $a+b+c$, we obtain $k = \frac{1}{2}$.
But if $a+b+c = 0$, then we have $b+c = -a$, etc.

$$\therefore k = \frac{a}{b+c} = -1.$$

(For example, if $a = b = 1$, $c = -2$, then $-1 = \frac{1}{1+(-2)} = \frac{1}{(-2)+1} = \frac{-2}{1+1}$.)

Alternate Solution.

$$\begin{aligned} k &= \frac{a}{b+c} = \frac{b}{c+a} = \frac{c}{a+b} \\ &= \frac{a+b+c}{2(a+b+c)} = \frac{1}{2} \quad (\text{if } a+b+c \neq 0). \end{aligned}$$

[2]This problem is a reminder to students of the *cardinal law of mathematics:*

"Thou shalt not divide by zero".

If $a+b+c=0$, then it is obvious that $k=-1$.

Exercise. Suppose u, v, w, x, y, z, k are real numbers such that

$$k=\frac{u}{x}=\frac{v}{y}=\frac{w}{z}.$$

(a) Show that, for any real numbers p, q, r,

$$k=\frac{pu+qv+rw}{px+qy+rz},$$

provided $px+qy+rz\neq 0$. If $px+qy+rz=0$, then $pu+qv+rw=0$.

(b) In addition, if $k\geq 0$, show that

$$\begin{aligned} k &= \sqrt{\frac{pu^2+qv^2+rw^2}{px^2+qy^2+rz^2}} \\ &= \sqrt{\frac{pvw+qwu+ruv}{pyz+qzx+rxy}}, \end{aligned}$$

provided none of the denominators is zero.

(c) Show that, for any real numbers a, b, c, d,

$$\frac{au+bx}{cu+dx}=\frac{av+by}{cv+dy}=\frac{aw+bz}{cw+dz},$$

provided none of the denominators is zero.

(d) Generalize.

35. $\displaystyle\frac{1}{3}=\frac{1}{5}+\frac{1}{9}+\frac{1}{45}.$

Without loss of generality, we may assume that $a<b<c$. Then because all the quantities are positive, we must have $3<a<9$.

(a) Suppose $a=5$, then

$$\frac{1}{b}+\frac{1}{c}=\frac{1}{3}-\frac{1}{5}=\frac{2}{15}.$$

If $b=7$, then $\frac{1}{c}=\frac{2}{15}-\frac{1}{7}<0$. If $b=9$, then $\frac{1}{c}=\frac{2}{15}-\frac{1}{9}=\frac{1}{45}$, $\therefore\ c=45$.

(b) Suppose $a=7$, then

$$\frac{1}{b}+\frac{1}{c}=\frac{1}{3}-\frac{1}{7}=\frac{4}{21}.$$

If $b=9$, then $\frac{1}{c}=\frac{4}{21}-\frac{1}{9}=\frac{5}{63}$, which is not a unit fraction. If $b\geq 11$, then $\frac{1}{c}\geq\frac{4}{21}-\frac{1}{11}=\frac{23}{21\cdot 11}>\frac{1}{b}$, which contradicts our assumption that $b<c$.

Thus, the only expression satisfying our conditions is

$$\frac{1}{3}=\frac{1}{5}+\frac{1}{9}+\frac{1}{45}.$$

Exercise. Show that, for any odd integer $n\geq 3$, there exist three distinct odd positive integers a, b, c such that

$$\frac{1}{n}=\frac{1}{a}+\frac{1}{b}+\frac{1}{c}.$$

36. $x = 3$, $y = 2$, $u = 8$, $v = 1$.

Note that

$$z + w = (x + 5) + (y - 1)\sqrt{2} = u + \sqrt{2}$$

implies

$$\begin{aligned} x + 5 &= u \quad \text{and} \quad y - 1 = 1. \quad \therefore\ y = 2. \\ zw &= (x - \sqrt{2})(5 + y\sqrt{2}) = (x - \sqrt{2})(5 + 2\sqrt{2}) \\ &= (5x - 4) + (2x - 5)\sqrt{2} = 11 + v\sqrt{2}. \\ \therefore\ 5x - 4 &= 11, \quad x = 3, \\ v &= 2x - 5 = 1, \quad u = x + 5 = 8. \end{aligned}$$

Cf. L.-s. Hahn, *Complex Numbers and Geometry,* Mathematical Association of America, Washington, D.C., 1994; Exercise 2, p.42.]

Question: Where did we use the assumption that x, y, u, v are rational?

37. (a) $u = 13, \quad v = 5$. (b) $x = 3, \quad y = 2$.

(a) Squaring both sides of the given equality

$$\sqrt{18 - 2\sqrt{65}} = \sqrt{u} - \sqrt{v},$$

we obtain

$$18 - 2\sqrt{65} = (u + v) - 2\sqrt{uv}.$$

$$\therefore\ u + v = 18, \quad uv = 65.$$

So u and v must be the roots of

$$t^2 - 18t + 65 = (t - 5)(t - 13) = 0. \quad \therefore\ t = 5,\ 13.$$

Because $\sqrt{18 - 2\sqrt{65}} > 0$, we have $u > v$, it follows that $u = 13$, $v = 5$.

(b) We work from inside. From our computation in Part (a), we know that if we can find two positive numbers m and n ($m > n$), whose sum and product are 3 and 2, respectively, then we have

$$\sqrt{3 - 2\sqrt{2}} = \sqrt{m} - \sqrt{n}.$$

It is easy to see that 2 and 1 satisfy the condition.

$$\therefore\ \sqrt{3 - 2\sqrt{2}} = \sqrt{2} - 1.$$

Repeating the same reasoning, we obtain

$$\begin{aligned} \sqrt{5 - 12\sqrt{3 - 2\sqrt{2}}} &= \sqrt{5 - 12(\sqrt{2} - 1)} \\ &= \sqrt{17 - 2\sqrt{72}} \end{aligned}$$

$$
\begin{aligned}
&= \sqrt{9}-\sqrt{8} = 3-2\sqrt{2}.\\
\therefore \sqrt{3+2\sqrt{5-12\sqrt{3-2\sqrt{2}}}} &= \sqrt{3+2(3-2\sqrt{2})}\\
&= \sqrt{9-2\sqrt{8}}\\
&= \sqrt{8}-1 = 2\sqrt{2}-1.\\
\therefore \sqrt{14+3\sqrt{3+2\sqrt{5-12\sqrt{3-2\sqrt{2}}}}} &= \sqrt{14+3(2\sqrt{2}-1)}\\
&= \sqrt{11+2\sqrt{18}}\\
&= \sqrt{9}+\sqrt{2} = 3+\sqrt{2}.
\end{aligned}
$$

38. $a = 42,\ b = 7,\ c = 3,\ p = 5.$

Squaring both sides of the given equality

$$\sqrt{94+a\sqrt{p}} = b + c\sqrt{p},$$

we obtain

$$
\begin{aligned}
94 + a\sqrt{p} &= (b^2 + c^2 p) + 2bc\sqrt{p}.\\
\therefore b^2 + c^2 p &= 94, \quad 2bc = a.
\end{aligned}
$$

Because $p \geq 2$, from the first equation, we obtain

$$
\begin{aligned}
c^2 p &< 94, \quad c^2 < \frac{94}{p} \leq \frac{94}{2} = 47.\\
\therefore 2 \leq c &\leq 6, \quad \text{and} \quad 1 \leq b \leq 9.
\end{aligned}
$$

Now, if c is an even number, then again from the equation $b^2 + c^2 p = 94$, we obtain

$$b^2 \equiv 94 \equiv 2 \pmod 4.$$

But no perfect square is congruent to $2 \pmod 4$. Hence c must be odd, which implies that $c = 3$ or $c = 5$.

Suppose $c = 5$. Then because $c^2 p = 25p < 94$, we have two cases.

(i) If $p = 2$, then $b^2 = 94 - c^2 p = 94 - 5^2 \cdot 2 = 44$, which is not a perfect square.

(ii) If $p = 3$, then $b^2 = 94 - 5^2 \cdot 3 = 19$, which is also not a perfect square.

Therefore, $c = 3$ is the only remaining possibility. In this case,

$$b^2 \equiv 94 \equiv 4 \pmod 9. \quad \therefore b \equiv \pm 2 \pmod 9.$$

So again we have two cases.

(i) If $b = 2$, then $2^2 + 3^2 p = 94$, $p = 10$, which is not a prime number.

(ii) If $b = 7$, then $7^2 + 3^2 p = 94$, $p = 5$, and $a = 2bc = 2 \times 7 \times 3 = 42$.

It is simple to verify that $a = 42$, $b = 7$, $c = 3$, and $p = 5$ satisfy the condition.

39. $p = q = \frac{3}{5}$, $r = -\frac{1}{5}$.

Substituting $n = 1, 2, 3$, into

$$\frac{1^4 + 2^4 + 3^4 + \cdots + (n-1)^4 + n^4}{1^2 + 2^2 + 3^2 + \cdots + (n-1)^2 + n^2} = pn^2 + qn + r,$$

we obtain, respectively,

$$p + q + r = 1, \quad 4p + 2q + r = \frac{17}{5}, \quad 9p + 3q + r = \frac{98}{14} = 7.$$

Solving this system of simultaneous equations, we obtain $p = q = \frac{3}{5}$, $r = -\frac{1}{5}$.

Remark. It can be shown, by mathematical induction, say, that with this choice of the coefficients p, q, r the equality actually holds for all positive integers n.

Exercise. Show that there exist coefficients p, q, r such that

$$\frac{1^5 + 2^5 + 3^5 + \cdots + (n-1)^5 + n^5}{1^3 + 2^3 + 3^3 + \cdots + (n-1)^3 + n^3} = pn^2 + qn + r,$$

for all positive integers n.

40. $x = 84$, $y = 85$.

The numbers in the third column are 1 more than those in the second column, so we set $y = x + 1$, and obtain

$$13^2 + x^2 = (x+1)^2.$$

Solving for x, we obtain $x = 84$, $y = 85$.

Alternate Solution I. The numbers in the second column

$$4,\ 12,\ 24,\ 40,\ 60,\ \cdots$$

are all multiples of 4. Hence dividing by 4, we obtain

$$1,\ 3,\ 6,\ 10,\ 15,\ \cdots$$

which are easily recognized as the triangle numbers, with the next term as 21. Thus $x = 4 \cdot 21 = 84$.

$$\therefore\ y^2 = 84^2 + 13^2 = 85^2.$$

Alternate Solution II. Rewriting the given equation, we obtain

$$13^2 = y^2 - x^2 = (y - x)(y + x).$$

But 13 is a prime (and x, y are positive integers), therefore

$$y + x = 13^2, \quad y - x = 1.$$

Thus $x = 84$, $y = 85$.

41. $x = 63, \quad y = 73.$

Clearly, the numbers in the first column

$$5,\ 7,\ 9,\ 11,\ 13,\ \cdots$$

are consecutive odd integers, and the differences of two neighboring terms in the third column

$$8-3,\ 15-8,\ 24-15,\ 35-24,\cdots$$

are identical to the numbers in the first column. Moreover, the differences of two neighboring terms in the column on the right

$$13-7,\ 21-13,\ 31-21,\ 43-31,\cdots$$

are consecutive even integers. Thus

$$15^2 + 15 \times 48 + 48^2 = 57^2, \quad 17^2 + 17 \times 63 + 63^2 = 73^2.$$

Alternate Solution. The numbers 5, 7, 9, 11, 13, $\cdots$ are clearly odd integers and can be represented by $2n+1$ $(n \geq 2)$. The numbers 3, 8, 15, 24, 35, $\cdots$ are one less than perfect squares, and so can be represented by $n^2 - 1$ $(n \geq 2)$. Therefore, in general, we have

$$\begin{aligned}(2n+1)^2 + (2n+1)(n^2-1) + (n^2-1)^2 &= n^4 + 2n^3 + 3n^2 + 2n + 1 \\ &= (n^2+n+1)^2.\end{aligned}$$

Setting $n = 8$, we obtain $17^2 + 17 \times 63 + 63^2 = 73^2$.

42. $(x, y) = (3612, 3613), (720, 725), (204, 221), (132, 157).$

Perhaps the easiest solution is obtained by setting $y = x+1$. Then the given equation can be rewritten as

$$85^2 + x^2 = (x+1)^2,$$

from which we obtain

$$x = \frac{85^2 - 1}{2} = 43 \cdot 84 = 3612, \quad \text{and } y = 3613.$$

To find all the solutions, we rewrite the given equation as

$$(y-x)(y+x) = 85^2 = 5^2 \cdot 17^2.$$

This gives four possibilities:

$$\begin{cases} y - x = 1 \\ y + x = 5^2 \cdot 17^2, \end{cases} \quad \begin{cases} y - x = 5 \\ y + x = 5 \cdot 17^2, \end{cases} \quad \begin{cases} y - x = 17 \\ y + x = 5^2 \cdot 17, \end{cases} \quad \begin{cases} y - x = 5^2 \\ y + x = 17^2. \end{cases}$$

Solving these four systems of simultaneous equations, we obtain

x	3612	720	204	132
y	3613	725	221	157

43. $x = 6$ or -2.

Newton solved $(x-3)(x+4) = x^2 + x - 12$, and Leibnitz solved $(x+1)(x-5) = x^2 - 4x - 5$. Hence the original equation must be $x^2 - 4x - 12 = (x-6)(x+2)$.

Alternate Solution. Assuming the leading coefficient is 1, the constant term of the quadratic equation must be $3 \cdot (-4) = -12$, and the coefficient of the first degree term must be $-(-1+5) = -4$. Hence the original equation must be $x^2 - 4x - 12 = (x-6)(x+2)$.

44. (a) $a = 2$, $b = -11$. (b) $c = 4$, $d = -1$.

(a)

$$\begin{aligned} x &= \frac{4+3\sqrt{3}}{2+\sqrt{3}} = \frac{(4+3\sqrt{3})(2-\sqrt{3})}{(2+\sqrt{3})(2-\sqrt{3})} = -1+2\sqrt{3}. \\ \therefore\ (x+1)^2 &= \left(2\sqrt{3}\right)^2 = 12. \quad x^2+2x-11=0. \end{aligned}$$

Exercise. Find the polynomial of the least degree whose coefficients are all integers, whose leading coefficient (i.e., the coefficient of the highest term) is 1, and which has $\sqrt{3}+\sqrt{5}$ as one of its roots.

(b) By long division, we obtain

$$\begin{aligned} f(x) &= x^4 + 2x^3 - 10x^2 + 4x - 10 \\ &= (x^2+2x-11)(x^2+1) + (2x+1). \end{aligned}$$

Because $\dfrac{4+3\sqrt{3}}{2+\sqrt{3}} = -1+2\sqrt{3}$ is a root of $x^2+2x-11$, we obtain

$$\begin{aligned} f\left(\frac{4+3\sqrt{3}}{2+\sqrt{3}}\right) &= 2(-1+2\sqrt{3})+1 \\ &= 4\sqrt{3}-1. \quad c=4, \quad d=-1. \end{aligned}$$

Exercise. Solve this problem using synthetic division.

45. The product is 1.

Computing the difference of the two given quadratic equations, we obtain $1996(x^2 - 1) = 0$. Clearly, $x = 1$ is not a root of either equation, but $x = -1$ is a common root.

$$\begin{aligned} 1997x^2 + 1998x + 1 &= (x+1)(1997x+1), \\ x^2 + 1998x + 1997 &= (x+1)(x+1997). \end{aligned}$$

Thus the other roots are $-\dfrac{1}{1997}$ and -1997, and their product is 1.

Remarks. (a) Note that the discriminant for either one of these quadratic equations is

$$\begin{aligned} b^2 - 4ac &= 1998^2 - 4 \cdot 1 \cdot 1997 = (1997+1)^2 - 4 \cdot 1997 \\ &= 1997^2 - 2 \cdot 1997 + 1 = (1997-1)^2 = 1996^2. \end{aligned}$$

(b) Suppose $ac \neq 0$. Then α is a root of $ax^2 + bx + c = 0$ if and only if $\dfrac{1}{\alpha}$ is a root of $cx^2 + bx + a = 0$. Moreover, this result generalizes to polynomials of any degree.

46. $k = 6$.

Let x_0 be the nonzero common root. Then

$$x_0^2 - 5x_0 + k = 0\,, \qquad x_0^2 - 9x_0 + 3k = 0\,.$$

Subtracting these two equalities, we obtain

$$4x_0 - 2k = 0\,, \qquad \therefore k = 2x_0\,.$$

Substituting this result into one of the equalities above, we obtain

$$x_0^2 - 5x_0 + 2x_0 = x_0(x_0 - 3) = 0, \qquad \therefore x_0 = 3 \quad (\because x_0 \neq 0).$$

Therefore, $k = 2x_0 = 2 \cdot 3 = 6$. In fact, when $k = 6$, the given quadratic polynomials are

$$x^2 - 5x + 6 = (x-3)(x-2), \quad x^2 - 9x + 18 = (x-3)(x-6),$$

and so they have $x = 3$ as the common root.

47. The roots of the given cubic polynomial are 1, 8 and -9.

Because the given quadratic polynomials divide the given cubic polynomial, the roots of the quadratic polynomials must be roots of the cubic polynomial too. However, a cubic polynomial can have only three roots, so the given two quadratic polynomials must have (at least) one (and only one, because $a \neq b$) root in common, say x_0. Then $x_0^2 + ax_0 + b = 0, \quad x_0^2 + bx_0 + a = 0$.

Subtracting these two equalities, we obtain $(a-b)x_0 + (b-a) = 0$. Because $a - b \neq 0$, this gives $x_0 = 1$.

Substituting this value into either one of the given two quadratic polynomials, we obtain $1 + a + b = 0$.

$$\begin{aligned} \therefore \qquad x^2 + ax + b &= x^2 - (1+b)x + b = (x-1)(x-b), \\ x^2 + bx + a &= x^2 - (1+a)x + a = (x-1)(x-a). \end{aligned}$$

Because the given cubic polynomial is divisible by these two quadratic polynomials, the three roots of the given cubic polynomial must be 1, a and b. As the leading coefficient of the given cubic polynomial is 1, we must have

$$x^3 + px^2 + qx + 72 = (x-1)(x-a)(x-b).$$

Comparing the coefficients of both sides, we obtain

$$\begin{aligned} p &= 1 + a + b = 0, \qquad q = ab + a + b, \qquad 72 = -ab. \\ \therefore &\ q = -72 - 1 = -73. \\ \therefore &\ x^3 + px^2 + qx + 72 = x^3 - 73x + 72. \end{aligned}$$

Because we already know that $x_0 = 1$ is a root of this cubic polynomial (by synthetic division, say), we obtain

$$x^3 - 73x + 72 = (x-1)(x^2 + x - 72) = (x-1)(x-8)(x+9).$$

Hence the three roots of the given cubic polynomial are 1, 8 and -9, and the given two quadratic polynomials are

$$x^2 - 9x + 8 \qquad \text{and} \qquad x^2 + 8x - 9.$$

Remark. The second half of the solution can be argued as follows: Because

$$x^3 + px^2 + qx + 72 = (x-1)(x-a)(x-b),$$

we obtain, by comparing the constant terms, $ab = -72$. This, together with $a+b = -1$, shows that a and b are the roots of the quadratic polynomial $t^2+t-72 = (t-8)(t+9)$. Therefore

$$\begin{cases} a = 8 \\ b = -9 \end{cases} \qquad \text{or} \qquad \begin{cases} a = -9 \\ b = 8. \end{cases}$$

In either case, the three roots of the given cubic polynomial must be 1, 8 and -9.

48. (a) $x = 6, -2, -3$. (b) $(-3, -2) \cup (6, \infty)$.

(a) For $x \geq 0$, the given equation $|x|x - 5x - 6 = 0$ becomes

$$x^2 - 5x - 6 = (x-6)(x+1) = 0. \quad x = 6 \text{ or } -1.$$

But $-1 < 0$, and so $x = 6$.

For $x < 0$, the given equation becomes

$$x^2 + 5x + 6 = (x+2)(x+3) = 0. \quad x = -2 \text{ or } -3.$$

(b) The set of all x that satisfy

$$x > 0 \text{ and } (x-6)(x+1) > 0$$

is the (infinite) interval $(6, \infty)$.

The set of all x that satisfy

$$x < 0 \text{ and } -(x+2)(x+3) > 0$$

is the interval $(-3, -2)$.

Hence $(-3, -2) \cup (6, \infty)$ is the set of all x that satisfy the given inequality.

Exercise. Sketch the graph of the function $f(x) = |x|x - 5x - 6$.

49. $p = -24$, $q = 144$.

Set

$$\begin{aligned} & x^4 + 2x^3 - 23x^2 + px + q \\ = & (x-a)^2 \cdot (x-b)^2 = (x^2 - 2ax + a^2)(x^2 - 2bx + b^2) \\ = & x^4 - 2(a+b)x^3 + (a^2 + 4ab + b^2)x^2 - 2ab(a+b)x + a^2b^2. \end{aligned}$$

Comparing the coefficients, we obtain

$$\begin{cases} a + b = -1, \\ a^2 + 4ab + b^2 = -23, \\ -2ab(a+b) = p, \\ a^2b^2 = q. \end{cases}$$

From the second equality, we obtain

$$\begin{aligned} -23 &= (a+b)^2 + 2ab = (-1)^2 + 2ab. \quad \therefore\ ab = -12. \\ p &= -2(-12)(-1) = -24, \quad q = (-12)^2 = 144. \end{aligned}$$

Because $a + b = -1$ and $ab = -12$, a and b are also the roots of

$$t^2 + t - 12 = (t+4)(t-3) = 0.$$

Hence the four roots of the original quartic equation are 3, 3, -4, -4.

50. $x = -13, -1, 8, 20$.

Suppose $x^2 - 7x - 4 = n^2$, then $x^2 - 7x - (n^2 + 4) = 0$.

$$\therefore\ x = \frac{7 \pm \sqrt{7^2 + 4(n^2+4)}}{2} = \frac{7 \pm \sqrt{4n^2 + 65}}{2}.$$

For x to be an integer, $4n^2+65$ must be the square of an odd integer. Set $4n^2+65 = p^2$. Then

$$(p + 2n)(p - 2n) = 65 = 5 \cdot 13.$$

Because $p + 2n > 0$, we have $p + 2n \geq p - 2n > 0$.

$$\therefore \begin{cases} p + 2n = 13, \\ p - 2n = 5; \end{cases} \quad \text{or} \quad \begin{cases} p + 2n = 65, \\ p - 2n = 1. \end{cases}$$

Solving these two systems of simultaneous equations, we obtain

$$p = 9,\ n = 2,\ x = 8, -1\ ; \text{ or } p = 33,\ n = 16,\ x = 20, -13.$$

Indeed,

$$\begin{aligned} f(8) &= 8^2 - 7 \cdot 8 - 4 = 4 = 2^2, \\ f(-1) &= (-1)^2 - 7 \cdot (-1) - 4 = 4 = 2^2, \\ f(20) &= 20^2 - 7 \cdot 20 - 4 = 256 = 16^2, \\ f(-13) &= (-13)^2 - 7 \cdot (-13) - 4 = 256 = 16^2. \end{aligned}$$

51. (a) $x^4 + \dfrac{1}{x^4} = y^4 - 4y^2 + 2.$ (b) $x^5 + \dfrac{1}{x^5} = y^5 - 5y^3 + 5y.$

(a)

$$\begin{aligned}
\left(x + \frac{1}{x}\right)^4 &= x^4 + 4x^2 + 6 + \frac{4}{x^2} + \frac{1}{x^4} \\
&= \left(x^4 + \frac{1}{x^4}\right) + 4\left(x^2 + \frac{1}{x^2}\right) + 6, \\
\therefore\ x^4 + \frac{1}{x^4} &= \left(x + \frac{1}{x}\right)^4 - 4\left(x^2 + \frac{1}{x^2}\right) - 6 \\
&= y^4 - 4(y^2 - 2) - 6 = y^4 - 4y^2 + 2.
\end{aligned}$$

(b)

$$\begin{aligned}
\left(x + \frac{1}{x}\right)^3 &= x^3 + 3x + \frac{3}{x} + \frac{1}{x^3} = \left(x^3 + \frac{1}{x^3}\right) + 3\left(x + \frac{1}{x}\right), \\
\therefore\ x^3 + \frac{1}{x^3} &= \left(x + \frac{1}{x}\right)^3 - 3\left(x + \frac{1}{x}\right) = y^3 - 3y. \\
\left(x + \frac{1}{x}\right)^5 &= x^5 + 5x^3 + 10x + \frac{10}{x} + \frac{5}{x^3} + \frac{1}{x^5} \\
&= \left(x^5 + \frac{1}{x^5}\right) + 5\left(x^3 + \frac{1}{x^3}\right) + 10\left(x + \frac{1}{x}\right), \\
\therefore\ x^5 + \frac{1}{x^5} &= \left(x + \frac{1}{x}\right)^5 - 5\left(x^3 + \frac{1}{x^3}\right) - 10\left(x + \frac{1}{x}\right) \\
&= y^5 - 5(y^3 - 3y) - 10y = y^5 - 5y^3 + 5y.
\end{aligned}$$

Alternate Solution.

(a) $x^4 + \dfrac{1}{x^4} = \left(x^2 + \dfrac{1}{x^2}\right)^2 - 2 = (y^2 - 2)^2 - 2 = y^4 - 4y^2 + 2.$

(b)

$$\begin{aligned}
x^3 + \frac{1}{x^3} &= \left(x^2 + \frac{1}{x^2}\right)\left(x + \frac{1}{x}\right) - \left(x + \frac{1}{x}\right) \\
&= (y^2 - 2)y - y = y^3 - 3y, \\
x^5 + \frac{1}{x^5} &= \left(x^3 + \frac{1}{x^3}\right)\left(x^2 + \frac{1}{x^2}\right) - \left(x + \frac{1}{x}\right) \\
&= (y^3 - 3y)(y^2 - 2) - y = y^5 - 5y^3 + 5y. \\
\text{Or,}\ x^5 + \frac{1}{x^5} &= \left(x^4 + \frac{1}{x^4}\right)\left(x + \frac{1}{x}\right) - \left(x^3 + \frac{1}{x^3}\right) \\
&= (y^4 - 4y^2 + 2)y - (y^3 - 3y) = y^5 - 5y^3 + 5y.
\end{aligned}$$

Exercise. Show that

(a) for every natural number n, $x^n + \dfrac{1}{x^n}$ is a polynomial in $y \left(= x + \dfrac{1}{x}\right)$ of degree n.

(b) the polynomial in Part (a) is an even function of y if n is an even integer, and it is an odd function of y if n is an odd integer.

52. (a) $y = 4$ or $y = -6$. (b) $x = 2 \pm \sqrt{3},\ -3 \pm 2\sqrt{2}$.

(a) Dividing both sides of

$$x^4 + 2x^3 - 22x^2 + 2x + 1 = 0$$

by x^2, we obtain

$$\left(x^2 + \frac{1}{x^2}\right) + 2\left(x + \frac{1}{x}\right) - 22 = 0.$$

Because $x^2 + \dfrac{1}{x^2} = y^2 - 2$, we obtain

$$(y^2 - 2) + 2y - 22 = 0, \quad \text{i.e.,}\ y^2 + 2y - 24 = 0.$$

$$(y-4)(y+6) = 0. \quad \therefore\ y = 4 \ \text{or}\ y = -6.$$

(b) If $y = 4$; i.e., $x + \dfrac{1}{x} = 4$, then

$$\therefore\ x^2 - 4x + 1 = 0, \quad x = 2 \pm \sqrt{3}.$$

If $y = -6$; i.e., $x + \dfrac{1}{x} = -6$, then

$$\therefore\ x^2 + 6x + 1 = 0, \quad x = -3 \pm 2\sqrt{2}.$$

Note that

$$\begin{aligned} x^4 + 2x^3 - 22x^2 + 2x + 1 &= (x^2 - 4x + 1)(x^2 + 6x + 1) \\ &= (x - 2 - \sqrt{3})(x - 2 + \sqrt{3})(x + 3 - 2\sqrt{2})(x + 3 + 2\sqrt{2}). \end{aligned}$$

Question: For what type of equation is this method applicable?

53. $-\dfrac{14}{15}$.

$$3x^2 + 4x + 5 = 0, \quad \therefore\ x = \frac{-2 \pm i\sqrt{11}}{3}.$$

$$\begin{aligned} \frac{\alpha}{\beta} + \frac{\beta}{\alpha} &= \frac{-2 + i\sqrt{11}}{-2 - i\sqrt{11}} + \frac{-2 - i\sqrt{11}}{-2 + i\sqrt{11}} \\ &= \frac{(2 - i\sqrt{11})^2 + (2 + i\sqrt{11})^2}{(2 + i\sqrt{11})(2 - i\sqrt{11})} = \frac{2\{2^2 + (i\sqrt{11})^2\}}{2^2 - (i\sqrt{11})^2} \\ &= \frac{2(4 - 11)}{4 + 11} = -\frac{14}{15}. \end{aligned}$$

Alternate Solution. Suppose α and β are the two roots of a quadratic equation

$$ax^2 + bx + c = 0 \quad (a \neq 0).$$

Then, by comparing the coefficients of both sides of $ax^2 + bx + c = a(x - \alpha)(x - \beta)$, we obtain

$$\alpha + \beta = -\frac{b}{a}, \quad \alpha\beta = \frac{c}{a}.$$

$$\begin{aligned} \therefore \frac{\alpha}{\beta} + \frac{\beta}{\alpha} &= \frac{\alpha^2 + \beta^2}{\alpha\beta} = \frac{(\alpha + \beta)^2 - 2\alpha\beta}{\alpha\beta} \\ &= \frac{\left(-\frac{b}{a}\right)^2 - 2\left(\frac{c}{a}\right)}{\left(\frac{c}{a}\right)} = \frac{b^2 - 2ac}{ac}. \end{aligned}$$

Thus, in our case we have

$$\frac{\alpha}{\beta} + \frac{\beta}{\alpha} = \frac{4^2 - 2 \cdot 3 \cdot 5}{3 \cdot 5} = -\frac{14}{15}.$$

Exercise. (a) Suppose α, β, γ are the three roots of a cubic equation

$$ax^3 + bx^2 + cx + d = 0 \quad (a \neq 0).$$

Express

$$\alpha + \beta + \gamma, \quad \beta\gamma + \gamma\alpha + \alpha\beta, \quad \alpha\beta\gamma$$

in terms of the coefficients a, b, c, d.

(b) Can you generalize to quartic equations? Quintic equations?

54. (a) $\alpha^2 + \beta^2 + \gamma^2 = 4$. (b) $\alpha^3 + \beta^3 + \gamma^3 = -9$.

Because α, β, γ are the roots of the cubic equation $x^3 - 2x + 3 = 0$, by the exercise at the end of the solution to the preceding problem, we have

$$\alpha + \beta + \gamma = 0, \quad \beta\gamma + \gamma\alpha + \alpha\beta = -2, \quad \alpha\beta\gamma = -3.$$

(a) $\alpha^2 + \beta^2 + \gamma^2 = (\alpha + \beta + \gamma)^2 - 2(\beta\gamma + \gamma\alpha + \alpha\beta) = -2(-2) = 4$.

(b) This problem is immediate if we remember the identity

$$\alpha^3 + \beta^3 + \gamma^3 - 3\alpha\beta\gamma = (\alpha + \beta + \gamma)(\alpha^2 + \beta^2 + \gamma^2 - \beta\gamma - \gamma\alpha - \gamma\beta).$$

Even if we do not remember this identity, we can still solve this problem as follows:

$$\begin{aligned} \alpha^3 + \beta^3 + \gamma^3 &= (\alpha + \beta + \gamma)(\alpha^2 + \beta^2 + \gamma^2) \\ &\quad - \{\alpha^2(\beta + \gamma) + \beta^2(\gamma + \alpha) + \gamma^2(\alpha + \beta)\} \\ &= (\alpha + \beta + \gamma)(\alpha^2 + \beta^2 + \gamma^2) \\ &\quad - \{\alpha(\alpha\beta + \gamma\alpha) + \beta(\beta\gamma + \alpha\beta) + \gamma(\gamma\alpha + \beta\gamma)\} \\ &= (\alpha + \beta + \gamma)(\alpha^2 + \beta^2 + \gamma^2) \\ &\quad - (\alpha + \beta + \gamma)(\beta\gamma + \gamma\alpha + \alpha\beta) + 3\alpha\beta\gamma \\ &= (\alpha + \beta + \gamma)(\alpha^2 + \beta^2 + \gamma^2 - \beta\gamma - \gamma\alpha - \alpha\beta) + 3\alpha\beta\gamma \\ &= 3\alpha\beta\gamma \quad (\because \alpha + \beta + \gamma = 0) \\ &= 3(-3) = -9. \end{aligned}$$

Alternate Solution. Because α is a solution of $x^3 - 2x + 3 = 0$, we have

$$\alpha^3 - 2\alpha + 3 = 0; \text{ i.e., } \alpha^3 = 2\alpha - 3.$$

Similarly, we have

$$\beta^3 = 2\beta - 3, \quad \gamma^3 = 2\gamma - 3.$$

Adding the last three equalities, we obtain

$$\alpha^3 + \beta^3 + \gamma^3 = 2(\alpha + \beta + \gamma) - 9 = -9.$$

Exercise. Find the values of $\alpha^4 + \beta^4 + \gamma^4$ and $\alpha^5 + \beta^5 + \gamma^5$.

55. (a) $s_1 s_2 - s_3$. $\quad s_1^3 - 3s_1 s_2 + 3s_3$.

(a)

$$\begin{aligned} (b+c)(c+a)(a+b) &= (s_1 - a)(s_1 - b)(s_1 - c) \\ &= s_1^3 - (a+b+c)s_1^2 + (bc + ca + ab)s_1 - abc \\ &= s_1^3 - s_1 s_1^2 + s_2 s_1 - s_3 = s_1 s_2 - s_3. \end{aligned}$$

(b)

$$\begin{aligned} a^3 + b^3 + c^3 &= (a+b+c)(a^2+b^2+c^2) - a^2(b+c) - b^2(c+a) - c^2(a+b) \\ &= s_1(s_1^2 - 2s_2) - a(ab+ca) - b(bc+ab) - c(ca+bc) \\ &= s_1^3 - 2s_1 s_2 - a(s_2 - bc) - b(s_2 - ca) - c(s_2 - ab) \\ &= s_1^3 - 2s_1 s_2 - (a+b+c)s_2 + 3abc \\ &= s_1^3 - 2s_1 s_2 - s_1 s_2 + 3s_3 = s_1^3 - 3s_1 s_2 + 3s_3. \end{aligned}$$

Alternate Solution. Suppose the problem is correct. Then because $a^3 + b^3 + c^3$ is homogeneous of degree 3, for suitable coefficients p, q, r, we must have

$$a^3 + b^3 + c^3 = ps_1^3 + qs_1 s_2 + rs_3.$$

Because this equality must be an identity, setting $b + c = 0$ (i.e., $c = -b$), we obtain

$$\begin{aligned} a^3 + b^3 + c^3 &= a^3 + b^3 + (-b)^3 = a^3 \\ &= pa^3 + qa(-b^2) + r(-ab^2) = pa^3 - (q+r)ab^2. \end{aligned}$$

Comparing the coefficients of both sides, we obtain

$$p = 1, \quad q + r = 0.$$

Setting $a = b = c = 1$, we obtain

$$\begin{aligned} a^3 + b^3 + c^3 &= 3 = p(3)^3 + q(3)(3) + r(1) \\ &= 27p + 8q = 27 + 8q \quad (\because\ p = 1,\ r = -q). \\ \therefore\ q &= -3. \\ a^3 + b^3 + c^3 &= s_1^3 - 3s_1 s_2 + 3s_3. \end{aligned}$$

It is simple to verify that this is indeed an identity.

Question: Solve Part (a) using the alternative method.

Exercise. Show that

$$(b-c)^2(c-a)^2(a-b)^2 = s_1^2s_2^2 + 18s_1s_2s_3 - 4s_2^3 - 4s_1^3s_3 - 27s_3^2.$$

In particular, if $s_1 = 0$, then

$$(b-c)^2(c-a)^2(a-b)^2 = -4s_2^3 - 27s_3^2.$$

56. (a) $uv = -2$. (b) $u^3 + v^3 = 45$.

(a) $2uv = (u+v)^2 - (u^2+v^2) = 3^2 - 13 = -4$. $\therefore\ uv = -2$.

(b) $u^3 + v^3 = (u+v)(u^2 - uv + v^2) = 3(13+2) = 45$.

Remark. The values of u and v are $\dfrac{3 \pm \sqrt{17}}{2}$.

Exercise. Suppose

$$\begin{cases} u+v+w &= 0, \\ u^2+v^2+w^2 &= 4, \\ u^3+v^3+w^3 &= -9. \end{cases}$$

Find the value of $u^4 + v^4 + w^4$.

57. (a) $\sin\theta \cdot \cos\theta = -\dfrac{1}{8}$. (b) $\sin^3\theta - \cos^3\theta = \dfrac{7\sqrt{5}}{16}$.

(a)

$$\begin{aligned} -2\sin\theta\cdot\cos\theta &= (\sin\theta - \cos\theta)^2 - (\sin^2\theta + \cos^2\theta) \\ &= \left(\frac{\sqrt{5}}{2}\right)^2 - 1 = \frac{1}{4}. \\ \therefore\ \sin\theta\cdot\cos\theta &= -\frac{1}{8}. \end{aligned}$$

(b)

$$\begin{aligned} \sin^3\theta - \cos^3\theta &= (\sin\theta - \cos\theta)(\sin^2\theta + \sin\theta\cos\theta + \cos^2\theta) \\ &= \left(\frac{\sqrt{5}}{2}\right)\left(1 - \frac{1}{8}\right) = \frac{7\sqrt{5}}{16}. \end{aligned}$$

58. (a) $\{-5, -1, 1, 3\}$. (b) $a = -16,\ b = -2$.

Because all the coefficients are integers, and the leading coefficient is 1, all the rational roots must be integers and are factors of the constant term 15; i.e., the four roots form a subset of $\{\pm 1, \pm 3, \pm 5, \pm 15\}$. The only combination of four integers from this set whose product is 15 (the constant term), and whose sum is -2 (the negative of the coefficient of x^3) is $\{-5, -1, 1, 3\}$.

$$\begin{aligned} (x+5)(x+1)(x-1)(x-3) &= x^4 + 2x^3 - 16x^2 - 2x + 15. \\ \therefore\ a &= -16, \quad b = -2. \end{aligned}$$

59. The value of the expression is 0.

Because the leading coefficient of the given quadratic equation is 3, we set

$$(x-2)(x-3)+(x-3)(x+1)+(x+1)(x-2)=3(x-\alpha)(x-\beta).$$

Substituting $x=-1$, 2, 3 into this equality, we obtain, respectively,

$$\begin{aligned}(-3)\cdot(-4) &= 3(-1-\alpha)(-1-\beta), \quad \therefore\ (\alpha+1)(\beta+1)=4.\\ (-1)\cdot(3) &= 3(2-\alpha)(2-\beta), \quad \therefore\ (\alpha-2)(\beta-2)=-1.\\ (4)\cdot(1) &= 3(3-\alpha)(3-\beta), \quad \therefore\ (\alpha-3)(\beta-3)=\frac{4}{3}.\end{aligned}$$

$$\therefore\ \frac{1}{(\alpha+1)(\beta+1)}+\frac{1}{(\alpha-2)(\beta-2)}+\frac{1}{(\alpha-3)(\beta-3)}=\frac{1}{4}-1+\frac{3}{4}=0.$$

Exercise. Given three distinct numbers a, b, c, let α and β be the roots of the quadratic equation

$$(x-b)(x-c)+(x-c)(x-a)+(x-a)(x-b)=0.$$

Show that

$$\frac{1}{(\alpha-a)(\beta-a)}+\frac{1}{(\alpha-b)(\beta-b)}+\frac{1}{(\alpha-c)(\beta-c)}=0.$$

60. (a) $\varphi=\arctan\frac{4}{3}\ (0<\varphi<\frac{\pi}{2})$. (b) $|k|\geq\sqrt{3}$.

(a) If there exists an angle φ such that

$$3\sin\theta+4\cos\theta=5\sin(\theta+\varphi) \quad \text{for all } \theta,$$

then, in particular, setting $\theta=0$, we obtain

$$4=5\sin\varphi, \quad \therefore\ \sin\varphi=\frac{4}{5}.$$

Similarly, setting $\theta=\frac{\pi}{2}$, we obtain

$$3=5\sin\left(\frac{\pi}{2}+\varphi\right)=5\cos\varphi, \quad \therefore\ \cos\varphi=\frac{3}{5}.$$

Because $\left(\frac{3}{5}\right)^2+\left(\frac{4}{5}\right)^2=1$, such an angle φ exists. Indeed, for $\varphi=\arctan\frac{4}{3}\ (0<\varphi<\frac{\pi}{2})$, we have

$$\begin{aligned}\cos\varphi &= \frac{3}{5}, \quad \text{and} \quad \sin\varphi=\frac{4}{5}.\\ \therefore\ 3\sin\theta+4\cos\theta &= 5\left(\frac{3}{5}\sin\theta+\frac{4}{5}\cos\theta\right)\\ &= 5(\sin\theta\cdot\cos\varphi+\cos\theta\cdot\sin\varphi)\\ &= 5\sin(\theta+\varphi).\end{aligned}$$

(b) Choose an angle φ such that $\cos\varphi = \dfrac{k}{\sqrt{k^2+2^2}}$ and $\sin\varphi = \dfrac{-2}{\sqrt{k^2+2^2}}$. Then

$$\begin{aligned} k\sin\theta - 2\cos\theta &= \sqrt{k^2+2^2}\left\{\frac{k}{\sqrt{k^2+2^2}}\sin\theta + \frac{-2}{\sqrt{k^2+2^2}}\cos\theta\right\} \\ &= \sqrt{k^2+4}\,\{\sin\theta\cdot\cos\varphi + \cos\theta\cdot\sin\varphi\} \\ &= \sqrt{k^2+4}\,\sin(\theta+\varphi). \end{aligned}$$

Hence the given equation has a solution if and only if

$$\sqrt{k^2+4} \geq \sqrt{7};\ \text{i.e.,}\ \ |k| \geq \sqrt{3}.$$

61. (a) $h = \dfrac{1}{2}$, $k = -\dfrac{1}{2}$. (b) The sum is $\dfrac{36}{55}$.

(a) Clearing the denominators in

$$\frac{1}{x^2-1} = \frac{h}{x-1} + \frac{k}{x+1},$$

we obtain

$$1 = h(x+1) + k(x-1).$$

Because this has to be an identity, it has to be true for all x. Substituting $x = 1$, we obtain $1 = 2h$. $\therefore\ h = \dfrac{1}{2}$. Similarly, substituting $x = -1$, we obtain $k = -\dfrac{1}{2}$. Alternately, comparing the coefficients of both sides of

$$1 = h(x+1) + k(x-1) = (h+k)x + (h-k),$$

we obtain

$$h + k = 0, \quad h - k = 1.$$

Solving this system of simultaneous equations, we obtain $h = \dfrac{1}{2}$, $k = -\dfrac{1}{2}$.

(b) By the identity in Part (a), we have

$$\begin{aligned} &\frac{1}{2^2-1} + \frac{1}{3^2-1} + \frac{1}{4^2-1} + \cdots + \frac{1}{10^2-1} \\ &= \frac{1}{2}\left\{\left(1-\frac{1}{3}\right) + \left(\frac{1}{2}-\frac{1}{4}\right) + \left(\frac{1}{3}-\frac{1}{5}\right)\right. \\ &\qquad \left. + \cdots + \left(\frac{1}{8}-\frac{1}{10}\right) + \left(\frac{1}{9}-\frac{1}{11}\right)\right\} \\ &= \frac{1}{2}\left\{\left(1+\frac{1}{2}\right) - \left(\frac{1}{10}+\frac{1}{11}\right)\right\} = \frac{1}{2}\left\{\frac{3}{2} - \frac{21}{110}\right\} \\ &= \frac{1}{2}\cdot\frac{3}{2}\left(1-\frac{7}{55}\right) = \frac{3}{4}\cdot\frac{48}{55} = \frac{3\cdot 12}{55} = \frac{36}{55}. \end{aligned}$$

62. (a) $a_{19} = 190$. (b) $\sum_{n=1}^{95} \frac{1}{a_n} = \frac{95}{48}$

(a) To find the general term a_n, let us carry out a little experiment.

$$\begin{aligned} a_1 &= \frac{1\cdot 2\cdot 3}{6} = 1; \\ a_1 + a_2 &= \frac{2\cdot 3\cdot 4}{6} = 4, \quad \therefore\ a_2 = 3; \\ a_1 + a_2 + a_3 &= \frac{3\cdot 4\cdot 5}{6} = 10, \quad \therefore\ a_3 = 6. \end{aligned}$$

From these computations, we see immediately that

$$\begin{aligned} a_n &= (a_1 + a_2 + \cdots + a_{n-1} + a_n) - (a_1 + a_2 + \cdots + a_{n-1}) \\ &= \frac{n(n+1)(n+2)}{6} - \frac{(n-1)n(n+1)}{6} \\ &= \frac{n(n+1)}{6}\{(n+2) - (n-1)\} = \frac{n(n+1)}{2}. \\ \therefore\ a_{19} &= \frac{19\cdot 20}{2} = 190. \end{aligned}$$

(b)

$$\begin{aligned} &\frac{1}{a_1} + \frac{1}{a_2} + \cdots + \frac{1}{a_{95}} \\ &= 2\left\{\frac{1}{1\cdot 2} + \frac{1}{2\cdot 3} + \cdots + \frac{1}{94\cdot 95} + \frac{1}{95\cdot 96}\right\} \\ &= 2\left\{\left(1 - \frac{1}{2}\right) + \left(\frac{1}{2} - \frac{1}{3}\right) + \cdots + \left(\frac{1}{94} - \frac{1}{95}\right) + \left(\frac{1}{95} - \frac{1}{96}\right)\right\} \\ &= 2\left(1 - \frac{1}{96}\right) = \frac{95}{48}. \end{aligned}$$

63. (a) $S(x+y) = S(x)\cdot C(y) + C(x)\cdot S(y)$.
(b) $C(x+y) = C(x)\cdot C(y) + S(x)\cdot S(y)$.
(c) $T(x+y) = \dfrac{T(x) + T(y)}{1 + T(x)\cdot T(y)}$.

(a)

$$\begin{aligned} S(x+y) &= \frac{2^{x+y} - 2^{-x-y}}{2} \\ &= \left(\frac{2^x - 2^{-x}}{2}\right)\cdot\left(\frac{2^y + 2^{-y}}{2}\right) + \left(\frac{2^x + 2^{-x}}{2}\right)\cdot\left(\frac{2^y - 2^{-y}}{2}\right) \\ &= S(x)\cdot C(y) + C(x)\cdot S(y). \end{aligned}$$

(b)

$$\begin{aligned} C(x+y) &= \frac{2^{x+y} + 2^{-x-y}}{2} \\ &= \left(\frac{2^x + 2^{-x}}{2}\right)\cdot\left(\frac{2^y + 2^{-y}}{2}\right) + \left(\frac{2^x - 2^{-x}}{2}\right)\cdot\left(\frac{2^y - 2^{-y}}{2}\right) \\ &= C(x)\cdot C(y) + S(x)\cdot S(y). \end{aligned}$$

(c)

$$\begin{aligned} T(x+y) &= \frac{S(x+y)}{C(x+y)} \\ &= \frac{S(x)\cdot C(y)+C(x)\cdot S(y)}{C(x)\cdot C(y)+S(x)\cdot S(y)} \\ &= \frac{\dfrac{S(x)}{C(x)}+\dfrac{S(y)}{C(y)}}{1+\dfrac{S(x)}{C(x)}\cdot\dfrac{S(y)}{C(y)}} = \frac{T(x)+T(y)}{1+T(x)\cdot T(y)}. \end{aligned}$$

64. $g(x) = \dfrac{1+x^2}{1+x^2+x^4}, \quad h(x) = \dfrac{x}{1+x^2+x^4}.$

Suppose $f(x) = g(x) + h(x)$, where g is even, and h is odd. Then

$$f(-x) = g(-x) + h(-x) = g(x) - h(x).$$

Adding and subtracting these two equalities, we obtain

$$g(x) = \frac{1}{2}\{f(x) + f(-x)\}, \quad h(x) = \frac{1}{2}\{f(x) - f(-x)\}.$$

Hence, if such a pair of functions exists, then they must be given by these formulas.

Conversely, given f, if we define a pair of functions g and h by these formulas, then it is clear that g is even and h is odd, and their sum is f.

This shows that such a decomposition is always possible and, furthermore, it is unique.

For $f(x) = \dfrac{1}{1-x+x^2}$, we have $f(-x) = \dfrac{1}{1+x+x^2}$.

$$\begin{aligned} \therefore\ g(x) &= \frac{1}{2}\left\{\frac{1}{1-x+x^2} + \frac{1}{1+x+x^2}\right\} = \frac{1+x^2}{(1-x+x^2)(1+x+x^2)}, \\ h(x) &= \frac{1}{2}\left\{\frac{1}{1-x+x^2} - \frac{1}{1+x+x^2}\right\} = \frac{x}{(1-x+x^2)(1+x+x^2)}. \end{aligned}$$

65. $g(x) = \dfrac{x}{\sqrt{(x^2+3)(1+3x^2)}}, \quad h(x) = \sqrt{\dfrac{1+3x^2}{x^2+3}}.$

Suppose, for all $x > 0$,

$$g\left(\frac{1}{x}\right) = g(x) > 0, \quad h\left(\frac{1}{x}\right) = \frac{1}{h(x)}, \text{ and } f(x) = g(x)\cdot h(x).$$

Then

$$f\left(\frac{1}{x}\right) = g\left(\frac{1}{x}\right)\cdot h\left(\frac{1}{x}\right) = g(x)\cdot\frac{1}{h(x)}.$$

From the last two equalities (and the condition $g(x) > 0$), we obtain

$$g(x) = \sqrt{f(x)\cdot f\left(\frac{1}{x}\right)}, \quad h(x) = \sqrt{\frac{f(x)}{f\left(\frac{1}{x}\right)}}.$$

By the computation above, we see that such a decomposition of f is not only possible, but also unique (provided $f(x) > 0$ for $x > 0$). In our case, we have

$$\begin{aligned} g(x) &= \left\{\frac{x}{x^2+3}\cdot\frac{x}{1+3x^2}\right\}^{1/2} = \frac{x}{\sqrt{(x^2+3)(1+3x^2)}}, \\ h(x) &= \left\{\frac{x}{x^2+3}\cdot\frac{1+3x^2}{x}\right\}^{1/2} = \sqrt{\frac{1+3x^2}{x^2+3}}. \end{aligned}$$

Exercise. (a) Give an example of a function h such that, for all $x > 0$,

$$h\left(\frac{1}{x}\right) = -h(x).$$

(b) Given a function f, show how to find a pair of functions g and h such that, for all $x > 0$,

$$g\left(\frac{1}{x}\right) = g(x), \quad h\left(\frac{1}{x}\right) = -h(x), \quad \text{and } f(x) = g(x) + h(x).$$

66. (a) $p = 4$. (b) $f(x) = \sin\dfrac{\pi x}{2}$.

(a) Let us perform an experiment. Because f is odd, we try only nonnegative values of x. Substituting various integer values for x into the functional equation $f(x+1) = f(-x+1)$, we obtain

$$\begin{aligned} f(0) &= 0 \quad \text{(because } f \text{ is an odd function)}, \\ f(1) &= a \quad \text{(say)}, \\ f(2) &= f(0) = 0, \\ f(3) &= f(-1) = -f(1) = -a, \\ f(4) &= f(-2) = -f(2) = 0, \\ f(5) &= f(-3) = -f(3) = a, \\ f(6) &= f(-4) = -f(4) = 0, \\ f(7) &= f(-5) = -f(5) = -a, \quad \text{etc.} \end{aligned}$$

Thus, we conjecture that f is periodic with period 4; i.e.,

$$f(x+4) = f(x) \quad \text{for all } x.$$

To prove this is actually the case, we compute as follows:

$$\begin{aligned} f(x+2) &= f((x+1)+1) = f(-(x+1)+1) \\ &= f(-x) = -f(x). \end{aligned}$$

Repeated application of this result gives

$$f(x+4) = f((x+2)+2) = -f(x+2) = f(x).$$

Hence f has period 4. The example in Part (b) shows that f does not have a smaller period.

(b) Clearly, $f(x) = \sin \dfrac{\pi x}{2}$ is an odd function, and

$$f(x+1) = \sin \frac{\pi(x+1)}{2} = \cos \frac{\pi x}{2}$$

is an even function. It is evident from the graph of $f(x) = \sin \dfrac{\pi x}{2}$ that f has no period smaller than 4.

67. (a) $f(x) = \dfrac{x}{x-1}$. (b) $g_n(x) = \left(\dfrac{x^2}{x-1}\right)^n$ (for any positive integer n) will do.

(a) Set $\dfrac{x}{x-1} = t$, and solve x in terms of t, we obtain

$$x = \frac{t}{t-1}. \quad \therefore\ f(t) = f\left(\frac{x}{x-1}\right) = x = \frac{t}{t-1}.$$

(b) Any nonconstant symmetric function of x and $\dfrac{x}{x-1}$ will do. For example,

$$\begin{aligned} g_n(x) &= \left\{x \cdot \left(\frac{x}{x-1}\right)\right\}^n = \left(\frac{x^2}{x-1}\right)^n, \quad \text{or} \\ g_n(x) &= x^n + \left(\frac{x}{x-1}\right)^n \quad \text{(where } n \text{ is a positive integer).} \end{aligned}$$

Naturally, there are infinitely many others.

Exercise. Find, if any, a nonconstant function h satisfying

$$h(x) = h\left(\frac{1}{1-x}\right) = h\left(\frac{x-1}{x}\right) \quad (x \neq 0,\ 1).$$

How many such functions are there?

68. $$f(x) = \frac{1}{2}\left\{(1-x) + \frac{1}{x} - \frac{x}{x-1}\right\} = \frac{1 - x^2 + x^3}{2x(1-x)}$$

is the unique function satisfying the given functional equation.

Let $t = \dfrac{1}{1-x}$, then

$$x = \frac{t-1}{t}, \quad \frac{x-1}{x} = \frac{1}{1-t}, \quad \frac{x}{x-1} = 1 - t,$$

and so the given functional equation

$$f\left(\frac{1}{1-x}\right) + f\left(\frac{x-1}{x}\right) = \frac{x}{x-1}$$

becomes

$$f(t) + f\left(\frac{1}{1-t}\right) = 1 - t.$$

Similarly, if we let $t = \dfrac{x-1}{x}$, then

$$x = \frac{1}{1-t}, \quad \frac{1}{1-x} = \frac{t-1}{t}, \quad \frac{x}{x-1} = \frac{1}{t},$$

and so we have

$$f\left(\frac{t-1}{t}\right) + f(t) = \frac{1}{t}.$$

Add the last two functional equations, then subtract the given functional equation; we obtain

$$f(t) = \frac{1}{2}\left\{(1-t) + \frac{1}{t} - \frac{t}{t-1}\right\} = \frac{1-t^2+t^3}{2t(1-t)}.$$

Therefore, if there is a function which satisfies the given functional equation, then it must be the unique funtion obtained above.

Conversely, it is simple to verify that this function satisfies the given functional equation. Suppose

$$f(t) = \frac{1}{2}\left\{(1-t) + \frac{1}{t} - \frac{t}{t-1}\right\}.$$

Then, setting $t = \dfrac{1}{1-x}$, we obtain

$$f\left(\frac{1}{1-x}\right) = \frac{1}{2}\left\{\frac{x}{x-1} + (1-x) - \frac{1}{x}\right\}.$$

Similarly, setting $t = \dfrac{x-1}{x}$, we obtain

$$f\left(\frac{x-1}{x}\right) = \frac{1}{2}\left\{\frac{1}{x} + \frac{x}{x-1} - (1-x)\right\}.$$

Adding the last two equalities, we obtain

$$f\left(\frac{1}{1-x}\right) + f\left(\frac{x-1}{x}\right) = \frac{x}{x-1}.$$

Exercise.

(a) Find a nonconstant function f satisfying the functional equation

$$\begin{aligned} f(x) &= f\left(\frac{1}{x}\right) = f(1-x) = f\left(\frac{1}{1-x}\right) \\ &= f\left(\frac{x-1}{x}\right) = f\left(\frac{x}{x-1}\right) \quad (x \neq 0,\, 1). \end{aligned}$$

How many such functions are there?

(b) Find a function f satisfying the functional equation

$$f\left(\frac{1}{x}\right)+f(1-x)+f\left(\frac{1}{1-x}\right) + f\left(\frac{x-1}{x}\right)+f\left(\frac{x}{x-1}\right)=x \quad (x\neq 0,\, 1).$$

Is such a function unique?

69. The multiplicity of the root $x=1$ is a positive even number.

Obviously, $x=1$ is a root of $g(x)-g(1)=0$. Let k be its multiplicity. Then $k\geq 1$, and there exists a rational function φ such that

$$g(x)-g(1)=(x-1)^k\cdot\varphi(x) \quad \text{and} \quad \varphi(1)\neq 0.$$

Substituting $\frac{1}{x}$ for x, and rewriting the result, we obtain

$$\begin{aligned} g\left(\frac{1}{x}\right)-g(1) &= \left(\frac{1}{x}-1\right)^k\cdot\varphi\left(\frac{1}{x}\right). \\ g(x)-g(1) &= \frac{(1-x)^k}{x^k}\varphi\left(\frac{1}{x}\right). \end{aligned}$$

Now comparing this result with the first equality, and substituting $x=1$ into the result, we obtain

$$\varphi(x)=\frac{(-1)^k}{x^k}\varphi\left(\frac{1}{x}\right). \quad (-1)^k=1 \quad (\because\ \varphi(1)\neq 0).$$

Therefore, k is a positive even integer.

Conversely, for an arbitrary positive even integer $k=2m$, let

$$g(x)=\left[\frac{(x-1)^2}{x}\right]^m.$$

Then

$$g\left(\frac{1}{x}\right)=g(x),$$

and the multiplicity of the root of

$$g(x)-g(1)=\left[\frac{(x-1)^2}{x}\right]^m-0=\frac{(x-1)^{2m}}{x^m}$$

at $x=1$ is $2m=k$.

Remark. The condition that g is a rational function is to ensure the 'smoothness' of the function at $x=1$—without which the conclusion is false.
Example. $g(x)=|\log x|$ $(x>0)$.

Question: What can be said about the multiplicity of the root of $g(x)-g(-1)=0$ at $x=-1$ (where g satisfies $g\left(\frac{1}{x}\right)=g(x)$)?

70.

(a) & (b) Suppose

$$p(x) = \sum_{k=0}^{2003} a_k x^k = a_0 + a_1 x + a_2 x^2 + \cdots + a_{2003} x^{2003}.$$

Let us consider a simpler case first. How do we express the following polynomials in terms of p?

$$\begin{aligned} q_0(x) &= \sum_{k=0}^{1001} a_{2k} x^{2k} = a_0 + a_2 x^2 + \cdots + a_{2002} x^{2002}, \\ q_1(x) &= \sum_{k=0}^{1001} a_{2k+1} x^{2k+1} = a_1 x + a_3 x^3 + \cdots + a_{2003} x^{2003}. \end{aligned}$$

It is obvious that q_0 is even, while q_1 is odd; i.e.,

$$\begin{aligned} q_0(-x) &= q_0(x), & q_1(-x) &= -q_1(x), \\ p(x) &= q_0(x) + q_1(x), & p(-x) &= q_0(x) - q_1(x), \\ q_0(x) &= \frac{p(x) + p(-x)}{2}, & q_1(x) &= \frac{p(x) - p(-x)}{2}. \end{aligned}$$

Moreover, 1 and -1 are the roots of $x^2 = 1$. This observation gives us a clue. Let ω be a primitive root of $x^3 = 1$ (i.e., $\omega^3 = 1$, $\omega \neq 1$). Then 1, ω, ω^2 are the three distinct roots of $x^3 = 1$. Note $1 + \omega + \omega^2 = 0$. Clearly,

$$p(x) = f_0(x) + f_1(x) + f_2(x),$$

where

$$f_2(x) = \sum_{k=0}^{667} a_{3k+2} x^{3k+2} = a_2 x^2 + a_5 x^5 + \cdots + a_{2003} x^{2003}.$$

We have

$$\begin{aligned} p(\omega x) &= f_0(\omega x) + f_1(\omega x) + f_2(\omega x) \\ &= f_0(x) + \omega f_1(x) + \omega^2 f_2(x), \\ p(\omega^2 x) &= f_0(x) + \omega^2 f_1(x) + \omega f_2(x). \\ \therefore\ p(x) + p(\omega x) + p(\omega^2 x) &= 3 f_0(x) + \left(1 + \omega + \omega^2\right) f_1(x) + \left(1 + \omega^2 + \omega\right) f_2(x) \\ &= 3 f_0(x). \\ f_0(x) &= \frac{p(x) + p(\omega x) + p(\omega^2 x)}{3}. \end{aligned}$$

Similarly,

$$\begin{aligned} f_1(x) &= \frac{p(x) + \omega^2 p(\omega x) + \omega p(\omega^2 x)}{3}, \\ f_2(x) &= \frac{p(x) + \omega p(\omega x) + \omega^2 p(\omega^2 x)}{3}. \end{aligned}$$

(c) It is now easy to obtain

$$g(x) = \frac{p(x) + p(ix) + p(-x) + p(-ix)}{4}.$$

71. If the chain of inequalities in 2^0 is shortened to merely

$$\left|\sum a_k\right| < \left|\sum a_k^2\right|,$$

then anyone in their right mind would probably choose two numbers, one positive, one negative. For a more drastic result, we can choose $a_1 = r$, $a_2 = -r$ so that

$$\sum a_k = 0, \quad \sum a_k^2 = 2r^2 \qquad (0 < r < 1).$$

Or, if the chain of inequalities is shortened to merely

$$\left|\sum a_k\right| < \left|\sum a_k^4\right| \qquad \text{or} \qquad \left|\sum a_k^2\right| < \left|\sum a_k^4\right|,$$

then the most likely candidate would be the quartet

$$b_1 = r, \quad b_2 = ir, \quad b_3 = -r, \quad b_4 = -ir \quad (0 < r < 1)$$

that gives

$$\sum b_k = \sum b_k^2 = 0, \quad \sum b_k^4 = 4r^4.$$

(Actually, we also have $\sum b_k^3 = 0$.) Ignoring the absolute valute for a moment, we note that 1 and -1 are just the two roots of $x^2 = 1$, and of course 1, i, -1, and $-i$ are precisely the four roots of $x^4 = 1$. Encouraged by this observation, suppose we choose three roots

$$c_1 = 1, \quad c_2 = \omega, \quad c_3 = \omega^2, \qquad \text{where} \quad \omega = \frac{1}{2}(-1 \pm i\sqrt{3}),$$

of $x^3 = 1$, then

$$\sum c_k = \sum c_k^2 = 0, \quad \sum c_k^3 = 3.$$

Hence if we set

$$a_1 = r, \quad a_2 = -r, \quad a_3 = r, \quad a_4 = r\omega, \quad a_5 = r\omega^2,$$

then we have

$$\sum a_k = 0, \quad \sum a_k^2 = 2r^2, \quad \sum a_k^3 = 3r^3.$$

Now all we need is to choose $r = \frac{4}{5}$, say (actually, arbitrary $\frac{2}{3} < r < 1$ will do), then we obtain

$$\left|\sum a_k\right| < \left|\sum a_k^2\right| < \left|\sum a_k^3\right|.$$

By now we know that the crucial observation is the following

Lemma For $m \geq 2$, suppose ω_m is a primitive root of $x^m = 1$ (i.e., $\omega_m^m = 1$, but $\omega_m^k \neq 1$ for $1 \leq k < m$). Then

$$\sum_{k=0}^{m-1} \omega_m^{k\ell} = \begin{cases} m & (\ell \equiv 0 \pmod{m}) \\ 0 & (\ell \not\equiv 0 \pmod{m}) \end{cases}$$

Proof. Substitute $x = \omega_m^\ell$ into the identity

$$x^m - 1 = (x-1)\left(x^{m-1} + x^{m-2} + \cdots + x + 1\right),$$

and observe that the left-hand side is always 0, while the first factor on the right is 0 if and only if $\ell \equiv 0 \pmod{m}$. □

Once we have the lemma above, we set

$$\begin{array}{lllll} a_1 = r, & a_2 = -r, & a_3 = r, & a_4 = r\omega_3, & a_5 = r\omega_3^2 \\ a_6 = r, & a_7 = ri, & a_8 = -r, & a_9 = -ri, & \\ a_{10} = r, & a_{11} = r\omega_5, & a_{12} = r\omega_5^2, & a_{13} = r\omega_5^3, & a_{14} = r\omega_5^4, \\ a_{15} = a_{10}, & a_{16} = a_{11}, & a_{17} = a_{12}, & a_{18} = a_{13}, & a_{19} = a_{14}, \end{array}$$

where $0 < r < 1$ is to be determined later. Then, we have

$$\sum_{k=1}^{19} a_k = 0, \quad \sum_{k=1}^{19} a_k^2 = 2r^2, \quad \sum_{k=1}^{19} a_k^3 = 3r^3, \quad \sum_{k=1}^{19} a_k^4 = 6r^4, \quad \sum_{k=1}^{19} a_k^5 = 10r^5.$$

Clearly, choosing $r = \frac{4}{5}$ satisfies all the conditions in the problem.

Remark. Note that our method is applicable even if the chain of inequalities in 2^0 is longer.

Exercise. Does there exist a sequence

$$\{a_k \,;\, k = 1, 2, \cdots, n\}$$

satisfying the following inequalities simultaneously?

1^0 $|a_k| < 1 \quad (k = 1, 2, \cdots, n)$.

2^0 $$\left|\sum_{k=1}^{n} a_k\right| < \frac{1}{2^2}\left|\sum_{k=1}^{n} a_k^2\right| < \frac{1}{3^3}\left|\sum_{k=1}^{n} a_k^3\right| < \frac{1}{4^4}\left|\sum_{k=1}^{n} a_k^4\right| < \frac{1}{5^5}\left|\sum_{k=1}^{n} a_k^5\right|.$$

72. *Solution I.* For each positive integer n, let t_n be the sum of the products in the right column. We begin by analyzing the case $n = 5$ in detail. For this purpose we divide the ordered sums by their initial terms as follows:

1^0 The ordered sums that start with 1.

$$\begin{array}{cr} 1+4 & 1\times 2\ (=1\times 2) \\ 1+3+1 & 1\times 2\times 1\ (=1\times 2) \\ 1+1+3 & 1\times 1\times 2\ (=1\times 2) \\ 1+2+2 & 1\times 3\times 3\ (=1\times 9) \\ 1+2+1+1 & 1\times 3\times 1\times 1\ (=1\times 3) \\ 1+1+2+1 & 1\times 1\times 3\times 1\ (=1\times 3) \\ 1+1+1+2 & 1\times 1\times 1\times 3\ (=1\times 3) \\ 1+1+1+1+1 & 1\times 1\times 1\times 1\times 1\ (=1\times 1) \end{array}$$

We see these ordered sums are precisely 1 plus all the ordered sums of 4. Therefore, the contribution of these ordered sums to t_5 is

$$1 \times (2+2+2+9+3+3+3+1) = t_4.$$

2^0 The ordered sums that start with 2.

$$\begin{array}{cr} 2+3 & 3\times 2\ (=3\times 2) \\ 2+2+1 & 3\times 3\times 1\ (=3\times 3) \\ 2+1+2 & 3\times 1\times 3\ (=3\times 3) \\ 2+1+1+1 & 3\times 1\times 1\times 1\ (=3\times 1) \end{array}$$

We see these ordered sums are precisely 2 plus all the ordered sums of 3. Therefore, the contribution of these ordered sums to t_5 is

$$3\times(2+3+3+1)=3t_3.$$

3^0 The ordered sums that start with 3.

$$\begin{array}{cr} 3+2 & 2\times 3\ (=2\times 3) \\ 3+1+1 & 2\times 1\times 1\ (=2\times 1) \end{array}$$

Again, these ordered sums are precisely 3 plus all the ordered sums of 2. Therefore, the contribution of these ordered sums to t_5 is $2\times(3+1)=2t_2$.

4^0 The only ordered sum that start with 4 is $4+1$ and its contribution to t_5 is $2\times 1=2t_1$.

5^0 The only ordered sum that start with 5 is 5 itself and its contribution to t_5 is 2.

Therefore, we obtain

$$t_5=t_4+3t_3+2(t_2+t_1+1).$$

Note that in the process we obtain

$$t_1=1=1^2,\ t_2=4=2^2,\ t_3=9=3^2,\ t_4=25=5^2,\ t_5=64=8^2.$$

These results strongly suggest that $t_k = f_{k+1}^2$, where f_k are Fibonacci numbers defined by

$$f_0=0,\quad f_1=1,\quad f_{k+1}=f_k+f_{k-1}.$$

By now it should be obvious that by dividing the ordered sums for $n+1$ into those that start with $1, 2, \cdots, n, n+1$, respectively, we obtain

$$t_{n+1}=t_n+3t_{n-1}+2\left(t_{n-2}+t_{n-3}+\cdots+t_2+t_1+1\right).$$

To prove that our conjecture $t_k=f_{k+1}^2$ is true for all k, we proceed by mathematical induction. We know the conjecture is true for the first few cases. Suppose it is true for $k\le n$. Then we have

$$\begin{aligned} t_{n+1} &= f_{n+1}^2+3f_n^2+2\left(f_{n-1}^2+f_{n-2}^2+\cdots+f_3^2+f_2^2+1\right) \\ &= f_{n+1}^2+f_n^2+2\left(f_n^2+f_{n-1}^2+f_{n-2}^2+\cdots+f_3^2+f_2^2+f_1^2\right). \end{aligned}$$

For our conjecture to be true for all k, it is necessary and sufficient that

$$t_{n+1}=f_{n+2}^2=(f_{n+1}+f_n)^2=f_{n+1}^2+f_n^2+2f_{n+1}f_n.$$

Comparing this expression with the one above, we need the equality

$$f_{n+1}f_n = f_n^2 + f_{n-1}^2 + f_{n-2}^2 + \cdots + f_3^2 + f_2^2 + f_1^2.$$

But this equality is easy to verify either geometrically or inductively. Hence our proof is complete.

Solution II. We continue to employ the notations used in Solution I. Once we obtain the expression

$$t_{n+1} = t_n + 3t_{n-1} + 2\left(t_{n-2} + t_{n-3} + \cdots + t_2 + t_1 + 1\right),$$

by computing the difference between this expression and the corresponding one for t_n, we obtain the recursive formula

$$t_{n+1} - 2t_n - 2t_{n-1} + t_{n-2} = 0.$$

Observe that $t_n = r^n$ satisfies this recursive formula if

$$r^{n+1} - 2r^n - 2r^{n-1} + r^{n-2} = 0 \quad \text{for all } n.$$

But all of these equations reduce to

$$r^3 - 2r^2 - 2r + 1 = 0,$$

which has three roots

$$r_1 = -1, \quad r_2 = \frac{3+\sqrt{5}}{2}, \quad r_3 = \frac{3-\sqrt{5}}{2}.$$

Actually, more is true: Not only does $t_n = r_i^n$ $(i = 1, 2, 3)$ satisfy the recursive formula

$$t_{n+1} - 2t_n - 2t_{n-1} + t_{n-2} = 0 \quad \text{for all } n,$$

but, even with arbitrary coefficients a, b, c, so does their "linear combination":

$$t_n = ar_1^n + br_2^n + cr_3^n.$$

We have to choose these three coefficients so that the resulting linear combination satisfies the initial conditions

$$t_1 = 1, \quad t_2 = 4, \quad t_3 = 9.$$

This requirement is attained by

$$a = \frac{2}{5}, \quad b = \frac{r_2}{5}, \quad c = \frac{r_3}{5}.$$

Once we have the first three terms, then the rest are determined by the recursive formula. We have

$$\begin{aligned} t_n &= \frac{1}{5}\left[2(-1)^n + r_2^{n+1} + r_3^{n+1}\right] \\ &= \left[\frac{1}{\sqrt{5}}\left\{\left(\frac{1+\sqrt{5}}{2}\right)^{n+1} - \left(\frac{1-\sqrt{5}}{2}\right)^{n+1}\right\}\right]^2 = f_{n+1}^2, \end{aligned}$$

by the Binet formula. We have used

$$r_2 = \left(\frac{1+\sqrt{5}}{2}\right)^2, \quad r_3 = \left(\frac{1-\sqrt{5}}{2}\right)^2, \quad r_2 \cdot r_3 = 1.$$

Solution III. We shall continue to use the notations employed in Solution II. Note that $r_2 + r_3 = 3$, and $r_2 \cdot r_3 = 1$. Let

$$f(x) = t_0 + t_1 x + t_2 x^2 + \cdots + t_n x^n + \cdots$$

be the generating function, where $t_0 = t_1 = 1$, $t_2 = 4$. Multiply the generating function by 1, $-2x$, $-2x^2$, x^3, respectively, and add the four resulting expressions, then because of the recursive formula

$$t_{n+1} - 2t_n - 2t_{n-1} + t_{n-2} = 0,$$

all the coefficients of the powers of x higher than 2 vanish, and we obtain

$$\begin{aligned}
(1 - 2x - 2x^2 + x^3) f(x) &= t_0 + (t_1 - 2t_0)x + (t_2 - 2t_1 - 2t_0)x^2 \\
\therefore\ f(x) &= \frac{1-x}{1 - 2x - 2x^2 + x^3} \\
&= \frac{1}{5}\left[\frac{2}{1+x} + \frac{1}{r_2 - x} + \frac{1}{r_3 - x}\right] \\
&= \frac{1}{5}\Big[2\left(1 - x + x^2 - x^3 + - \cdots\right) \\
&\qquad + \frac{1}{r_2}\left(1 + r_3 x + r_3^2 x^2 + \cdots\right) \\
&\qquad + \frac{1}{r_3}\left(1 + r_2 x + r_2^2 x^2 + \cdots\right)\Big] \\
\therefore\ t_n &= \frac{1}{5}\left[(-1)^n \cdot 2 + r_2^{n+1} + r_3^{n+1}\right] \\
&= \left[\frac{1}{\sqrt{5}}\left\{\left(\frac{1+\sqrt{5}}{2}\right)^{n+1} - \left(\frac{1-\sqrt{5}}{2}\right)^{n+1}\right\}\right]^2 = f_{n+1}^2
\end{aligned}$$

as in Solution II.

Chapter 2.2 Geometry and Combinatorics

101. The sum of the 12 angles is 4π.

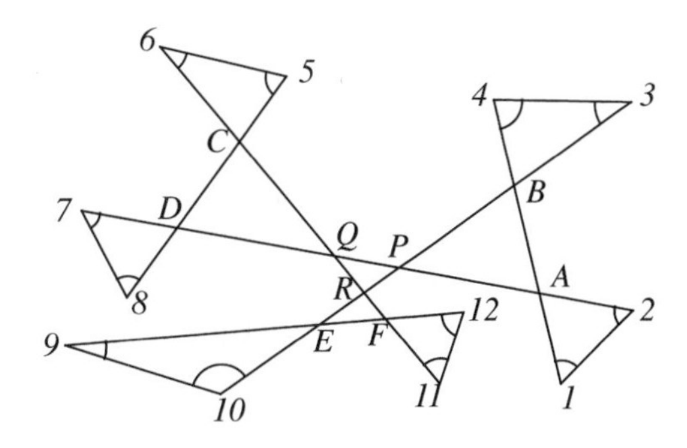

Figure 101(a)

Label as in Figure 101(a). Then

$$\begin{aligned} \angle 1 + \angle 2 &= \pi - \angle PAB, \\ \angle 3 + \angle 4 &= \pi - \angle PBA. \\ \therefore\ \angle 1 + \angle 2 + \angle 3 + \angle 4 &= 2\pi - (\angle PAB + \angle PBA) \\ &= 2\pi - (\pi - \angle APB) \\ &= \pi + \angle APB = \pi + \angle RPQ. \end{aligned}$$

Similarly,

$$\begin{aligned} \angle 5 + \angle 6 + \angle 7 + \angle 8 &= \pi + \angle PQR, \\ \angle 9 + \angle 10 + \angle 11 + \angle 12 &= \pi + \angle QRP. \\ \therefore\ \angle 1 + \angle 2 + \cdots + \angle 12 &= 3\pi + (\angle RPQ + \angle PQR + \angle QRP) \\ &= 3\pi + \pi = 4\pi. \end{aligned}$$

102. $a = 1,\ b = \frac{1}{2},\ c = -1.$

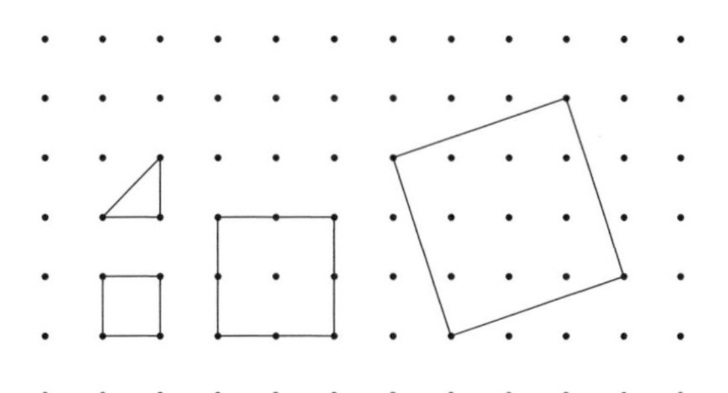

Figure 102(a)

Choosing a unit square ($p = 0,\ q = 4$), we obtain $4b + c = 1$. Choosing an isosceles right triangle half the size of a unit square ($p = 0,\ q = 3$), we obtain $3b + c = \frac{1}{2}$.

Solving these two equations for b and c, we obtain $b = \frac{1}{2}$, $c = -1$. Now, if we choose a 2×2 square ($p = 1$, $q = 8$), we obtain $a + 8b + c = 4$. $\therefore\ a = 1$.

$$\therefore \text{ Area } = p + \frac{1}{2}q - 1.$$

103. No, there exits no lattice equilateral triangle.

Suppose $\triangle ABC$ is an equilateral triangle, then its area is

$$\begin{aligned} \frac{1}{2} \cdot \overline{AB} \cdot \overline{AC} \cdot \sin\frac{\pi}{3} &= \frac{1}{2} \cdot \overline{AB}^2 \cdot \frac{\sqrt{3}}{2} \\ &= \frac{\sqrt{3}}{4}\left\{(a_1 - b_1)^2 + (a_2 - b_2)^2\right\}, \end{aligned}$$

which is irrational if $A = (a_1, a_2)$ and $B = (b_1, b_2)$ are lattice points (unless the equilateral triangle degenerates to a point). But according to the result of the preceding problem, the area of a simple lattice polygon must be a rational number. The contradiction establishes that it is impossible to have an equilateral triangle whose vertices are at lattice points.

Remark. Using the formula that the area of a triangle whose vertices are at (a_1, b_1), (a_2, b_2), (a_3, b_3), is given by the absolute value of the determinant

$$\frac{1}{2}\begin{vmatrix} a_1 & b_1 & 1 \\ a_2 & b_2 & 1 \\ a_3 & b_3 & 1 \end{vmatrix},$$

we see immediately that if all 3 vertices are at lattice points, then the area must be a rational number.

Trigonometric Solution. Without loss of generality, we may assume that one of the vertices is at the origin O, and $P = (a, b)$, $Q = (c, d)$ are other two vertices. Let φ and θ be the angles formed by OP and OQ with the positive x-axis, respectively. Again without loss of generality, we may assume that $\varphi < \theta$. Then

$$\begin{aligned} \sqrt{3} &= \tan\frac{\pi}{3} = \tan(\theta - \varphi) = \frac{\tan\theta - \tan\varphi}{1 + \tan\theta \cdot \tan\varphi} \\ &= \frac{\frac{d}{c} - \frac{b}{a}}{1 + \frac{d}{c} \cdot \frac{b}{a}} = \frac{ad - bc}{ac + bd}. \end{aligned}$$

The left-hand side is an irrational number, while the right-hand side is a rational number (if P and Q are lattice points). The contradiction establishes the non-existence of a lattice equilateral triangle.

Brute Force Solution. Given two vertices (a, b) and (c, d) of an equilateral triangle, by a straightforward computation, the third vertex must be one of the two points

$$\left(\frac{a + c \pm \sqrt{3}(b - d)}{2}, \frac{b + d \mp \sqrt{3}(a - c)}{2}\right).$$

For this third vertex to be a lattice point (assuming that a, b, c, d are integers), we must have $b - d = 0$ and $a - c = 0$. Namely, the equilateral triangle must degenerate to a point.

Slick Solution. Given a lattice triangle ABC, let $CPQR$ be the smallest rectangle that contains $\triangle ABC$ and whose sides are parallel to the coordinate axes (Figure 103).

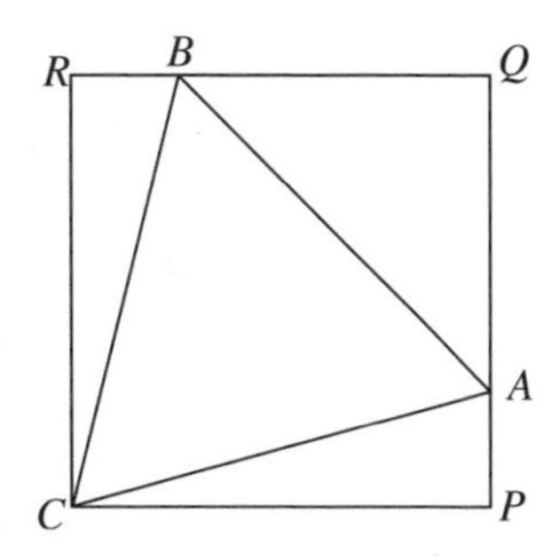

Figure 103

Then P, Q, R are all lattice points, and so the area of the rectangle $CPQR$, as well as those of $\triangle CPA$, $\triangle AQB$, $\triangle BRC$ are all rational. It follows that the area of $\triangle ABC$ must also be rational. But the area of an equilateral triangle is $\frac{1}{2} \cdot \overline{BC}^2 \cdot \sin\frac{\pi}{6} = \frac{\sqrt{3}}{4} \cdot \overline{BC}^2$. Now $\overline{BC}^2$ is rational (when the end points are lattice points), which implies that the area of a lattice equilateral triangle must be irrational. The contradiction proves the non-existence of a lattice equilateral triangle.

Remark. (a) The only regular polygons having all the vertices at lattice points in a plane must be squares.
(b) The only regular polygons having all the vertices at lattice points in a (3-dimentional) space must be squares, equilateral triangles or regular hexagons. (Can you give examples of such equilateral triangles and regular hexagons?)

104. (a) $(-16, 0)$, $(9, 0)$, $(0, 12)$. (b) $(0, 0)$, $(3, 3)$, $(-4, 4)$.
Naturally, there exist infinitely many solutions to all (a), (b) and (c).

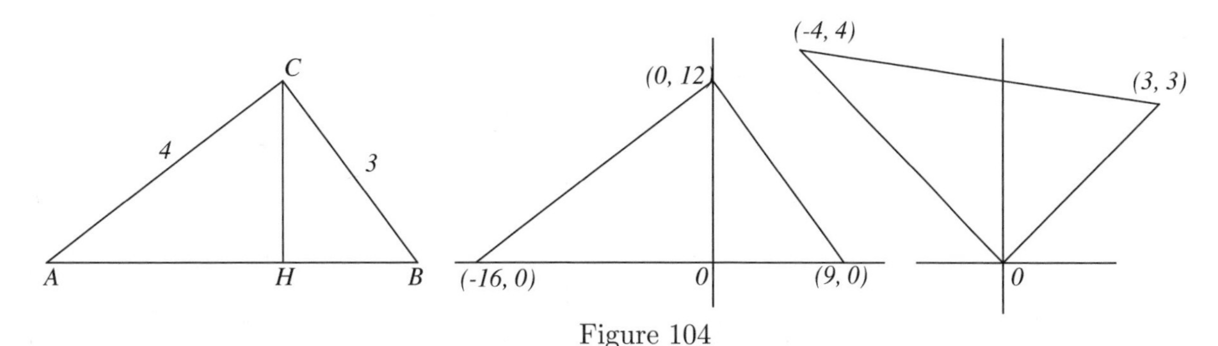

Figure 104

(a) In $\triangle ABC$, suppose

$$\overline{BC} = 3, \quad \overline{CA} = 4, \quad \overline{AB} = 5.$$

Let H be the foot of the altitude from vertex C to side AB. Then because

$$[ABC] = \frac{1}{2}\overline{AB} \cdot \overline{CH} = \frac{1}{2}\overline{BC} \cdot \overline{CA},$$

we have

$$5\overline{CH} = 3 \cdot 4, \quad \therefore\ \overline{CH} = \frac{12}{5}.$$

Moreover,

$$\begin{aligned} \triangle ABC &\sim \triangle ACH \sim \triangle CBH. \\ \therefore\ \frac{\overline{AC}}{\overline{AB}} &= \frac{\overline{AH}}{\overline{AC}}, \quad \overline{AH} = \frac{\overline{AC}^2}{\overline{AB}} = \frac{4^2}{5} = \frac{16}{5}. \\ \overline{BH} &= \overline{AB} - \overline{AH} = 5 - \frac{16}{5} = \frac{9}{5}. \end{aligned}$$

If follows that if we magnify by the factor 5, then all the lengths becomes integers. Hence placing side AB on the x-axis with H at the origin, we see that the triangle with vertices at

$$(-16,\ 0), \quad (9,\ 0), \quad (0,\ 12)$$

is similar to the 3-4-5 right triangle, and exactly one (the hypotenuse) of its sides is parallel to the coordinate axes.

(b) Note that the lines $y = x$ and $y = -x$ are perpendicular to each other, hence for any a and b, the triangle with vertices at

$$(0,\ 0), \quad (a,\ a), \quad (-b, b)$$

is a right triangle. Clearly, none of its sides is parallel to the coordinate axes unless $a = \pm b$, which is impossible for a Pythagorean triangle. (Why?) To make this triangle similar to the 3-4-5 right triangle, simply choose $a = 3,\ b = 4$.

Solution. By Ben Conlee (an eighth grader).

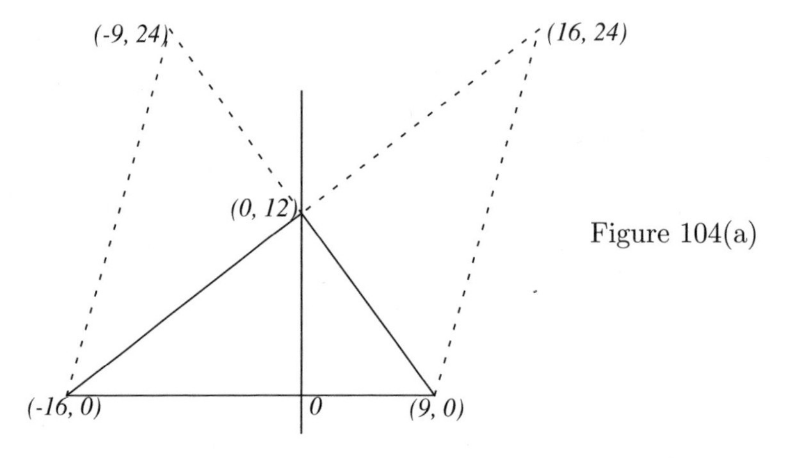

Figure 104(a)

Once we solved (a), then the solution of (b) can be obtained by reflecting the right triangle obtained in (a) with respect to one of its legs. For example, the right triangle with vertices at

$$(9,\ 0), \quad (0,\ 12), \quad (-16,\ 0)$$

is a solution to (a), so reflecting this right triangle with respect to its two legs, we obtain the right triangles with vertices at

$$(9,\,0),\quad (0,\,12),\quad (16,\,24)\quad \text{and}\quad (-9,\,24),\quad (0,\,12),\quad (-16,\,0)$$

that are two solutions to (b).

Alternate Solution. Let $P(x,\,y)$ be an arbitrary lattice point not on the coordinate axes (i.e., $xy \neq 0$.) Its image under the rotation around the origin O by $\frac{\pi}{2}$ is $Q(-y,\,x)$, and $OP \perp OQ$, $\overline{OP} = \overline{OQ}$. Extend one of them, say OP, three times, and the other four times, we reach points $P'(3x,\,3y)$ and $Q'(-4y,\,4x)$. Then $\triangle OP'Q'$ is a lattice triangle similar to the 3-4-5 right triangle.

(a) The hypotenuse of $\triangle OP'Q'$ is parallel to the x-axis if and only if $3y = 4x$. Setting $x = 3,\ y = 4$, we obtain a triangle with vertices at

$$(0,\,0),\quad (9,\,12),\quad (-16,\,12)$$

that is similar to the 3-4-5 right triangle and exactly one of its sides is parallel to the coordinate axes.

(b) None of the sides of $\triangle OP'Q'$ is parallel to the coordinate axes if and only if

$$3x \neq -4y,\quad 3y \neq 4x.$$

Choosing $x = 1,\ y = 2$, we obtain a triangle with vertices at

$$(0,\,0),\quad (3,\,6),\quad (-8,\,4)$$

that is similar to the 3-4-5 right triangle and none of its sides is parallel to the coordinate axes.

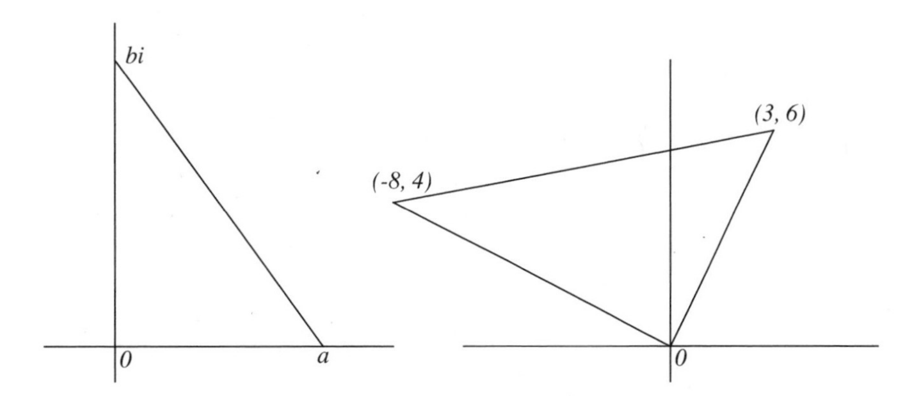

Figure 104(b)

Remark. Using complex numbers,[3] the last solution can be presented as follows: Let 0, a, and bi be the vertices of a right triangle (in the complex plane) with leg lengths a and b. (In our case, $a = 3,\ b = 4$, or $a = 4,\ b = 3$.) We rotate (around the origin) and magnify simultaneously by multiplying by $x + yi$, where x and y are integers. Then the origin remains fixed, but the vertex a moves to $a(x + yi) = ax + ayi$, and the vertex bi moves to $bi(x + yi) = -by + bxi$.

[3] See, for example, L.-s. Hahn: *Complex Numbers and Geometry,* Mathematical Association of America, 1994.

(a) The hypotenuse of the image triangle is parallel to the x-axis if and only if $ay = bx$, which is satisfied by choosing $x = a$ and $y = b$. (Naturally, there are other choices.) Then the new vertices are

$$0, \quad a^2 + abi, \quad -b^2 + abi\,; \text{ i.e., } (0, 0), \quad (a^2, ab), \quad (-b^2, ab).$$

For $a = 3,\ b = 4$, this gives three vertices as

$$(0, 0), (9, 12), (-16, 12).$$

(b) None of the sides of the image triangle is parallel to the coordinate axes if and only if

$$ay \neq bx, \quad ax \neq by.$$

For example, setting $x = y = 1$, we obtain the new vertices as

$$0, \quad a + ai, \quad -b + bi\,; \text{ i.e., } (0,0), \quad (a, a), \quad (-b, b).$$

If we choose $x = 1,\ y = 2$, then the new vertices are

$$0, \quad a + 2ai, \quad -2b + bi\,; \text{ i.e., } (0, 0), \quad (a, 2a), \quad (-2b, b).$$

For $a = 3,\ b = 4$, this gives three vertices as

$$(0, 0), \quad (3, 6), \quad (-8, 4).$$

Naturally, there are many other choices.

(c) Note that our methods for Parts (a) and (b) work for any Pythagorean triangle.

105. (a) 40. (b) 30.

(a) There are 20 squares with area 1, 12 squares with area 4, 6 squares with area 9, and 2 squares with area 16.
Hence there are $20 + 12 + 6 + 2 = 40$ squares.

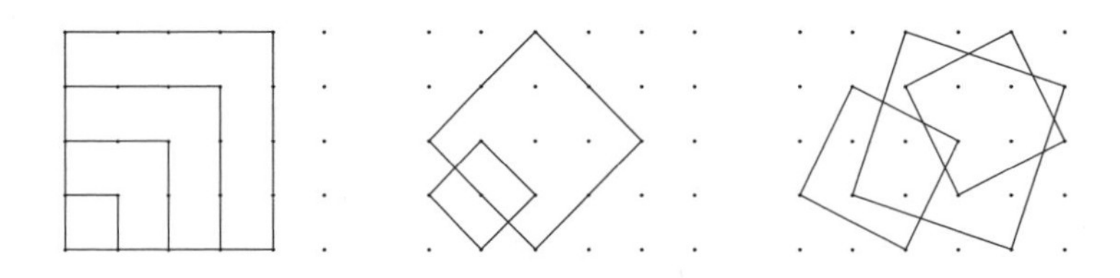

Figure 105(a)

(b) There are 12 squares with area 2, 12 squares with area 5, 2 squares with area 8, and 4 squares with area 10.

$$\therefore\ 12 + 12 + 2 + 4 = 30.$$

Question: What if the word "square" is replaced by "rectangle" in the problem?

106. The sum of the two angles is $\frac{\pi}{4}$ $(= 45°)$.

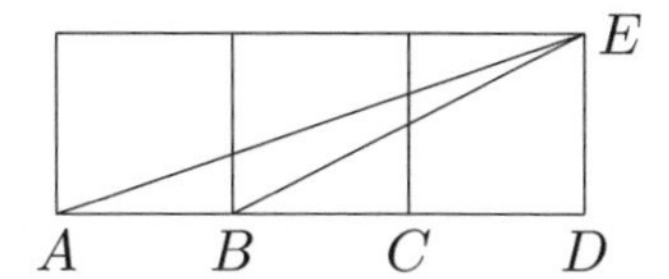

Figure 106

Let $\theta = \angle DAE$, $\varphi = \angle DBE$. Then

$$\tan(\theta + \varphi) = \frac{\tan\theta + \tan\varphi}{1 - \tan\theta \cdot \tan\varphi} = \frac{\frac{1}{3} + \frac{1}{2}}{1 - \frac{1}{3} \cdot \frac{1}{2}} = 1.$$

Because $\angle DAE + \angle DBE$ is clearly an angle in the first quadrant, we obtain

$$\angle DAE + \angle DBE = \frac{\pi}{4}.$$

Alternate Solution. A picture is worth a thousand words.

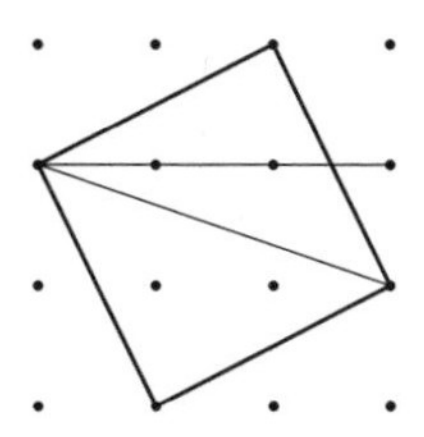

Figure 106(a)

Question: Can you generalize?

107. $\angle CPD = \frac{5\pi}{6}$.

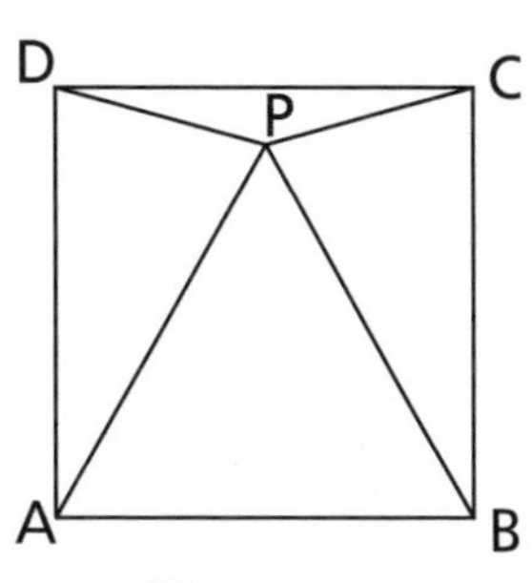

Figure 107

Because $\triangle BCP$ is an isosceles triangle and $\angle CBP = \frac{\pi}{6}$,

$$\angle BCP = \frac{5\pi}{12} \quad \text{and} \quad \angle PCD = \frac{\pi}{12}.$$

Similarly, $\angle PDC = \frac{\pi}{12}$. $\therefore$ $\angle CPD = \frac{5\pi}{6}$.

Question: What if P is outside the square (but the rest of the statement of the problem remains the same)?

108. $\overline{EF} = 6\ cm.$

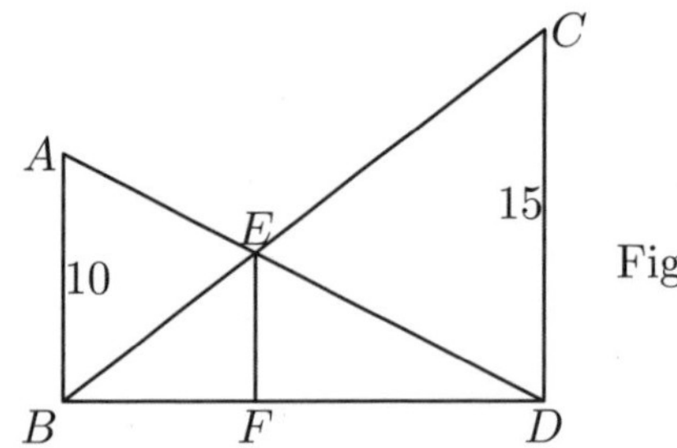

Figure 108

Let $\overline{BF} = a$, $\overline{FD} = b$, $\overline{EF} = x$. Then, because $\Delta DEF \sim \Delta DAB$,

$$\frac{\overline{EF}}{\overline{AB}} = \frac{\overline{FD}}{\overline{BD}}; \quad \text{i.e.,} \quad \frac{x}{10} = \frac{b}{a+b}.$$

Similarly,

$$\begin{aligned}
\frac{\overline{EF}}{\overline{CD}} &= \frac{\overline{BF}}{\overline{BD}}; \quad \text{i.e.,} \quad \frac{x}{15} = \frac{a}{a+b}. \\
\therefore \frac{x}{10} + \frac{x}{15} &= \frac{b}{a+b} + \frac{a}{a+b} = 1. \quad \therefore x\left(\frac{1}{10} + \frac{1}{15}\right) = 1. \\
\overline{EF} &= x = \frac{10 \cdot 15}{10+15} = 6.
\end{aligned}$$

Remark. Actually, AB, CD, EF do not have to be perpendicular to BD. They just need to be mutually parallel.

109. 34 units.

A picture is worth a thousand words.

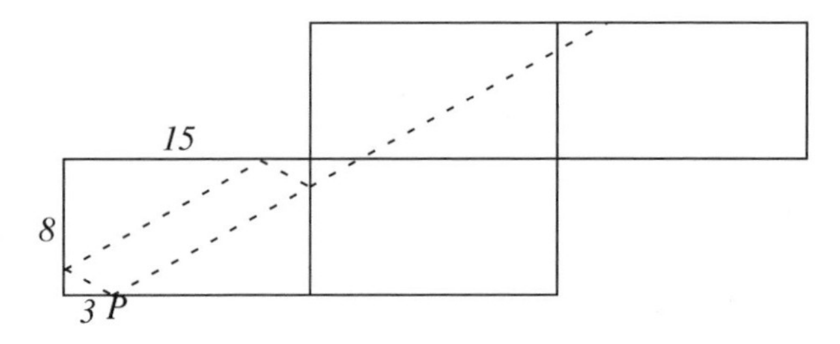

Figure 109(a)

$$\begin{aligned}
\sqrt{(2 \times 15)^2 + (2 \times 8)^2} &= 2\sqrt{15^2 + 8^2} \\
&= 2\sqrt{289} = 2 \cdot 17 = 34.
\end{aligned}$$

Note that the length of the path is independent of the starting point as long as it is not at one of the corners.

110. $[ABCD] = 35\ cm^2$.

A picture is worth a thousand words.

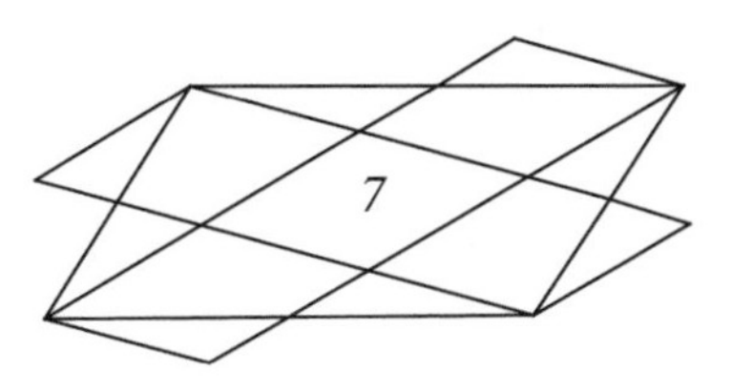

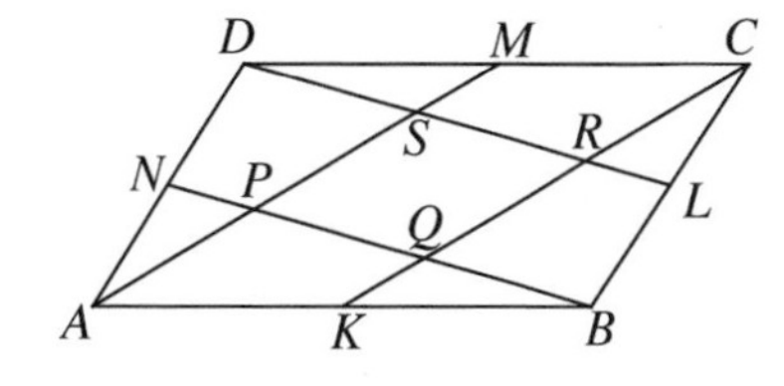

Figure 110(a)

Alternate Solution. Let $[ABCD]$ denote the area of the quadrangle $ABCD$, with similar notations for other quadrangles. Then clearly,

$$[AKCM] = \frac{1}{2}[ABCD].$$

Now, in ΔABP,

$$\overline{AK} = \overline{KB}, \quad KQ \parallel AP, \qquad \therefore \overline{KQ} = \frac{1}{2}\,\overline{AP} = \frac{1}{2}\,\overline{RC}.$$

Similarly, in ΔCBQ,

$$\overline{CL} = \overline{LB}, \quad LR \parallel BQ, \qquad \therefore \overline{RC} = \overline{QR}.$$

It follows that

$$\begin{aligned} \overline{QR} : \overline{KC} &= \overline{QR} : (\overline{KQ} + \overline{QR} + \overline{RC}) = 1 : \frac{5}{2} = 2 : 5. \\ \therefore [PQRS] &= \frac{2}{5}[AKCM] = \frac{2}{5} \cdot \frac{1}{2}[ABCD] = \frac{1}{5}[ABCD]. \end{aligned}$$

Hence, if $[PQRS] = 7cm^2$, then $[ABCD] = 35cm^2$.

111. The volume is $72\ cm^3$.

Let x, y, z be length, width and height. Then

$$yz = 12, \quad zx = 18, \quad xy = 24.$$

Multiplying these three equations together, we obtain

$$\begin{aligned} (xyz)^2 &= (yz) \cdot (zx) \cdot (xy) = 12 \cdot 18 \cdot 24 \\ &= (2^2 \cdot 3) \cdot (2 \cdot 3^2) \cdot (2^3 \cdot 3) = 2^6 \cdot 3^4. \\ \therefore\ xyz &= 2^3 \cdot 3^2 = 72\ cm^3. \end{aligned}$$

112. The area is $32\ cm^2$.

It is simple to see that the areas of $\triangle ABE$, $\triangle AFD$, and $\triangle CFE$ are 1/4, 1/4, and 1/8, respectively, of that of parallelogram $ABCD$. Hence the area of $\triangle AEF$ is $1 - 1/4 - 1/4 - 1/8 = 3/8$ of that of parallelogram $ABCD$. Therefore,

$$[ABCD] = \frac{8}{3}[AEF] = \frac{8}{3} \cdot 12 = 32\ cm^2.$$

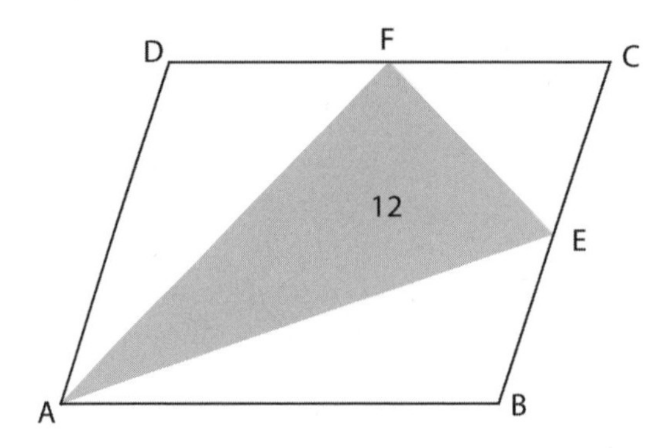

Figure 112

113. $[ABCD] = 74\ cm^2$.

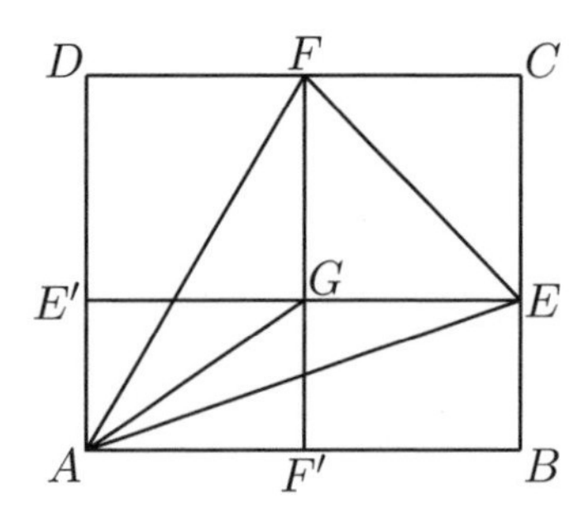

Figure 113(a)

Through E and F, draw lines EE' and FF' parallel to the sides of the rectangle, as in Figure 113(a). Let G be the intersection of EE' and FF'. Then

$$\begin{aligned} [ABCD] &= [AF'GE'] + [BEGF'] + [CFGE] + [DE'GF] \\ &= [AF'GE'] + 2\{[GAE] + [GEF] + [GFA]\} \\ &= [AF'GE'] + 2[AEF] = 4 \cdot 6 + 2 \cdot 25 = 74\ cm^2. \end{aligned}$$

Question: What if E or F coincides with C?

114. $[ABCD] = 49\ cm^2$.

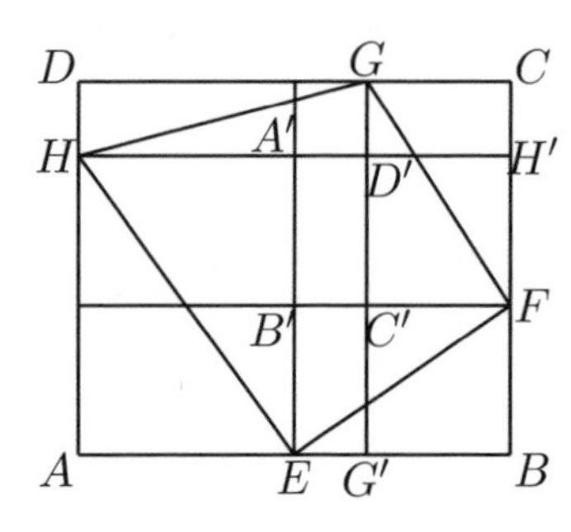

Figure 114(a)

Let A' be the intersection of HH' and the line through E parallel to AD; B' the intersection of lines through E, F parallel to BC, AB, respectively; C' the intersection of GG' and the line through F parallel to CD; D' the intersection of GG' and HH'. Then

$$\begin{aligned}[ABCD] &= [AEA'H] + [BFB'E] + [CGC'F] \\ &\quad + [DHD'G] + [A'B'C'D'] \\ &= 2\{[A'HE] + [B'EF] + [C'FG] + [D'GH]\} \\ &\quad + [A'B'C'D'] \\ &= 2\{[EFGH] - [A'B'C'D']\} + [A'B'C'D'] \\ &= 2[EFGH] - [A'B'C'D'] \\ &= 2 \cdot 32 - 3 \cdot 5 = 49\ cm^2.\end{aligned}$$

Question: What if G' and H' coincide with B and C, respectively? (Or, E and F coincide with A and B, respectively?)

115. The lengths of the two legs are 20 and 21.

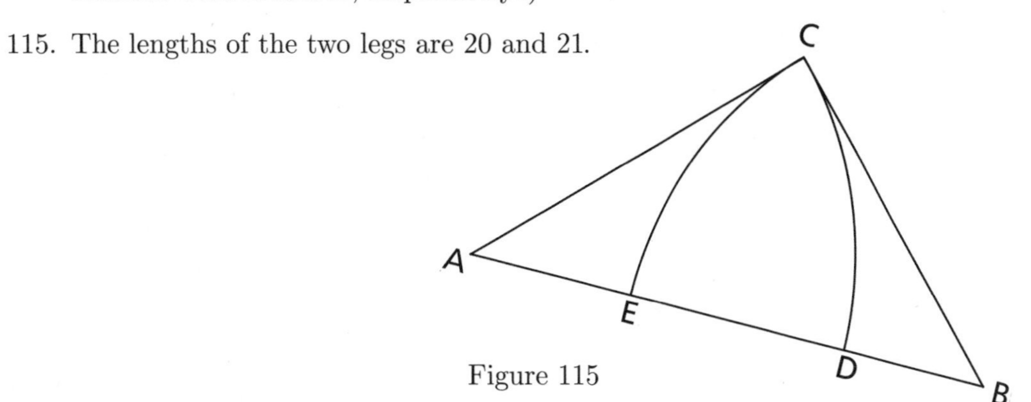

Figure 115

Let $a = \overline{BC}$, $b = \overline{CA}$. Then,

$$\begin{aligned} a + b &= \overline{AB} + \overline{DE} = 29 + 12 = 41. \\ a^2 + b^2 &= 29^2. \\ \therefore\ 2ab &= (a+b)^2 - (a^2+b^2) = 41^2 - 29^2 = 840. \\ (a-b)^2 &= (a^2+b^2) - 2ab = 29^2 - 840 = 1. \\ \therefore\ a - b &= \pm 1. \end{aligned}$$

Thus

$$\begin{cases} a = 21, \\ b = 20; \end{cases} \quad \text{or} \quad \begin{cases} a = 20, \\ b = 21. \end{cases}$$

116. (a) 120 cm^2. (b) 30 cm.

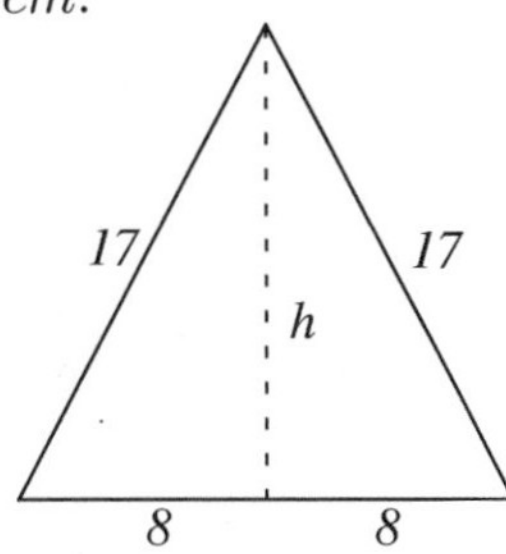

Figure 116

(a) Let h be the height on the side with length 16 cm. Then

$$\begin{aligned} h^2 &= 17^2 - \left(\frac{16}{2}\right)^2 = 289 - 64 = 225 = 15^2, \\ h &= 15. \\ \text{Area} &= \frac{1}{2} \cdot 16 \cdot 15 = 120\ cm^2. \end{aligned}$$

Alternate Solution. Use Heron's formula:

$$\text{Area} = \sqrt{s(s-a)(s-b)(s-b)}, \quad s = \frac{1}{2}(a+b+c).$$

In our case, $s = \frac{1}{2}(17 + 17 + 16) = 25.$

$$\begin{aligned} \text{Area} &= \sqrt{25(25-17)(25-17)(25-16)} \\ &= \sqrt{5^2 \cdot 2^6 \cdot 3^2} = 5 \cdot 2^3 \cdot 3 = 120\ cm. \end{aligned}$$

(b) Let the length of the third side be $2x$. Then

$$x\sqrt{17^2 - x^2} = 120. \quad x^2(17^2 - x^2) = 120^2.$$

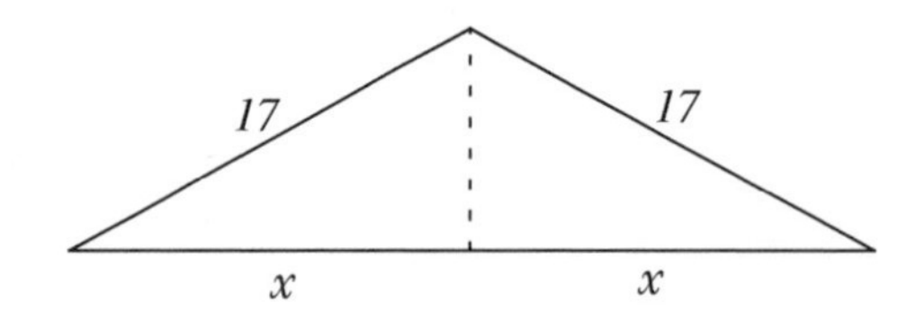

Figure 116(a)

Because we already know that $x = 8$ is one of the solutions of this equation (from Part (a)), we can rewrite the last equation as

$$x^4 - 17^2x^2 + 8^2 \cdot 15^2 = (x^2 - 8^2)(x^2 - 15^2) = 0.$$

Because we are looking for a solution $x > 8$, we must have $x = 15$; i.e., the third side is 30 cm.

Alternate Solution I. Let x be the same as above, and y the height on it. Then

$$x^2 + y^2 = 17^2, \quad xy = 120.$$

$$\begin{aligned} (x+y)^2 &= x^2 + y^2 + 2xy = 17^2 + 2 \cdot 120 = 289 + 240 = 529. \\ \therefore\ x + y &= 23. \\ (x-y)^2 &= x^2 + y^2 - 2xy = 289 - 240 = 49. \\ \therefore\ x - y &= \pm 7. \end{aligned}$$

It follows that

$$x = 15,\ y = 8 \ \text{ or } x = 8,\ y = 15.$$

The second solution gives the acute isosceles triangle of Part (a). Hence the desired length of the third side of the obtuse triangle must be $2x = 30$ cm.

Alternate Solution II. With geometric observations, no computation is needed at all! If we let one of the sides with length 17 cm serve as a base, then, because the acute and obtuse triangles have the same area, they must have the same height, and the two triangles fit together, forming a right triangle. (Why?)

Now the length $2x$ of the third side of the obtuse triangle we are looking for is twice the height of the acute triangle we computed in Part (a). Answer: 30 cm.

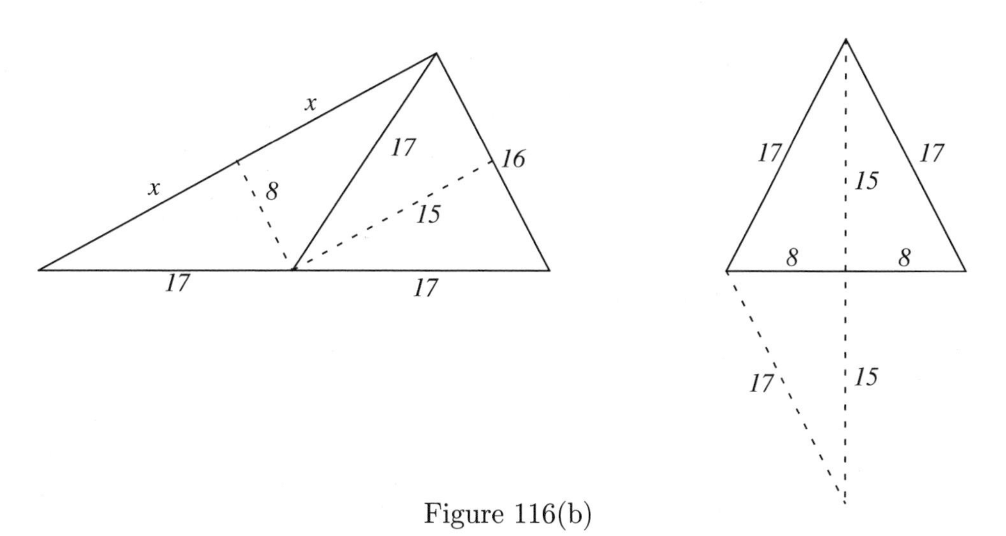

Figure 116(b)

Alternate Solution III. Cut the acute isosceles triangle in Part (a) into two congruent right triangles along a height (perpendicular), and then put them together to form an obtuse isosceles triangle. We see immediately that the length of the third side is 30 cm. (That there is only one obtuse triangle satisfying the required condition is clear.)

117. $\overline{BC} = 11$ cm.

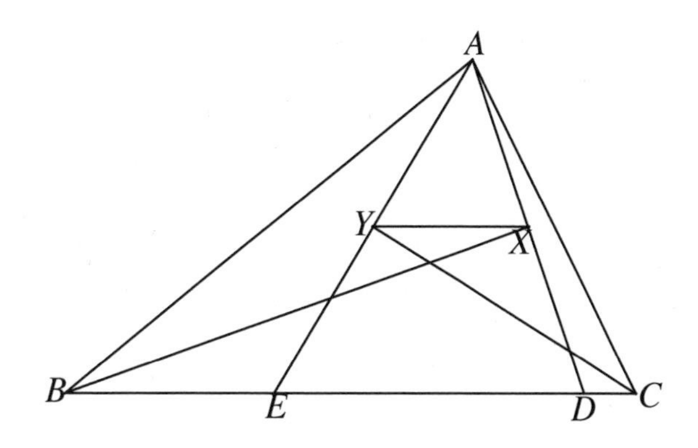

Figure 117

Let D, E be the intersections of BC and the extensions of AX and AY, respectively. Then

$$\begin{aligned} \triangle BDX &\cong \triangle BAX, \quad \overline{BD} = \overline{BA}; \\ \triangle CEY &\cong \triangle CAY, \quad \overline{CE} = \overline{CA}, \end{aligned}$$

and so X, Y are the midpoints of AD and AE, respectively.

$$\begin{aligned} \therefore\ \overline{XY} &= \frac{1}{2}\overline{DE} = \frac{1}{2}\{\overline{BD} + \overline{CE} - \overline{BC}\} \\ &= \frac{1}{2}\{\overline{BA} + \overline{CA} - \overline{BC}\}. \\ \therefore\ \overline{BC} &= \overline{BA} + \overline{CA} - 2\overline{XY} = 10 + 7 - 2 \cdot 3 = 11\ (cm). \end{aligned}$$

Exercise. In $\triangle ABC$, let X and Y be the feet of the perpendiculars from vertex A to the bisectors of exterior angles B and C, respectively. Show that

$$\overline{XY} = \frac{1}{2}\{\overline{BC} + \overline{CA} + \overline{AB}\}.$$

118. $[DEF] = 18\ cm^2$.

The following self-explanatory figure indicates the relation $[AEF] = 8[ABC]$.

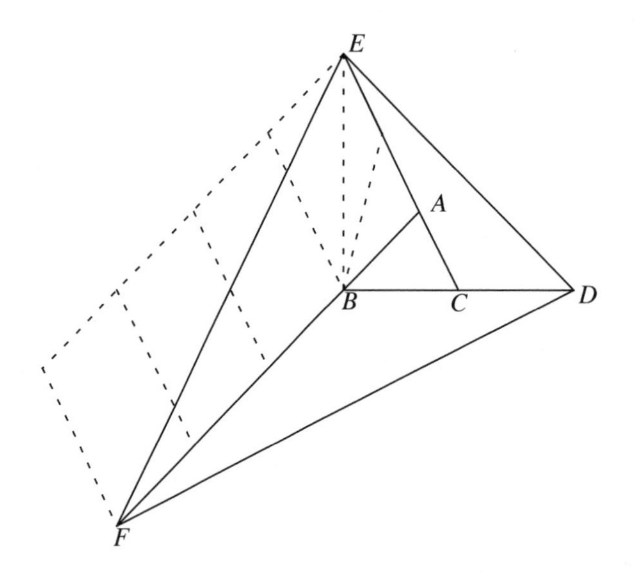

Figure 118(a)

By the same reasoning,

$$\begin{aligned} [BFD] &= 6[ABC], \quad [CDE] = 3[ABC]. \\ \therefore\ [DEF] &= [ABC] + [AEF] + [BFD] + [CDE] \\ &= (1 + 8 + 6 + 3)[ABC] = 18\ cm^2. \end{aligned}$$

119. $[ABC] = 18\ cm^2$.

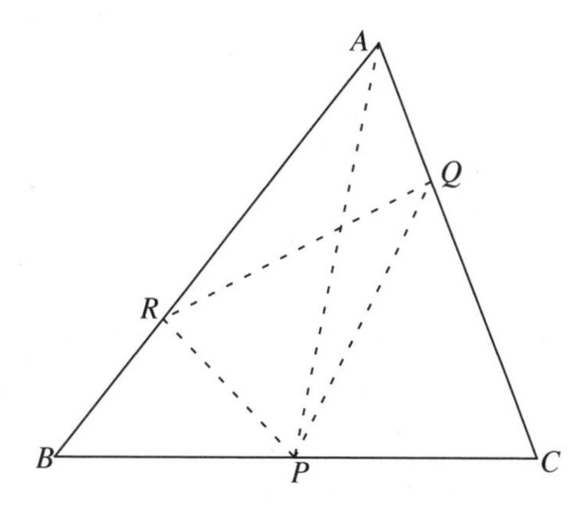

Figure 119(a)

Because $\overline{BP} = \frac{1}{2}\overline{BC}$, we have $[ABP] = \frac{1}{2}[ABC]$. Furthermore, because $\overline{RB} = \frac{1}{3}\overline{AB}$, we have $[BPR] = \frac{1}{3}[ABP]$.

$$\therefore\ [BPR] = \frac{1}{3}\cdot\frac{1}{2}[ABC] = \frac{1}{6}k,$$

where $k = [ABC]$. Similarly,

$$\begin{aligned}
[CQP] &= \frac{1}{2}\cdot\frac{2}{3}[ABC] = \frac{1}{3}k,\\
[ARQ] &= \frac{2}{3}\cdot\frac{1}{3}[ABC] = \frac{2}{9}k.\\
[PQR] &= [ABC] - \{[ARQ] + [BPR] + [CQR]\}\\
&= \left\{1 - \left(\frac{2}{9} + \frac{1}{6} + \frac{1}{3}\right)\right\} k = \frac{5}{18}k.
\end{aligned}$$

Hence for $[ARQ]$, $[BPR]$, $[CQP]$, $[PQR]$ to be integers, $k = [ABC]$ must be a multiple of 18. Moreover, for them to be consecutive integers, $[ABC] = 18$ is the only choice, and in this case,

$$[BPR] = 3, \quad [ARQ] = 4, \quad [PQR] = 5, \quad [CQP] = 6$$

constituting 4 consecutive integers.

Alternate Solution. It is easy to see that, of the four small triangles, $\triangle BPR$ is the smallest and $\triangle CQP$ is the largest. Moreover, because these two triangles have the same base, but the height of the latter is twice that of the former, we have

$$[CQP] = 2[BPR].$$

Setting $[BPR] = n$, we obtain $[CQP] = n + 3$, and so $2n = n + 3$; i.e., $n = 3$.

$$\therefore\ [ABC] = 3 + 4 + 5 + 6 = 18.$$

120. (a) $\overline{BC} = 8\ cm$. (b) $[ABC] = 10\sqrt{3}\ cm^2$.

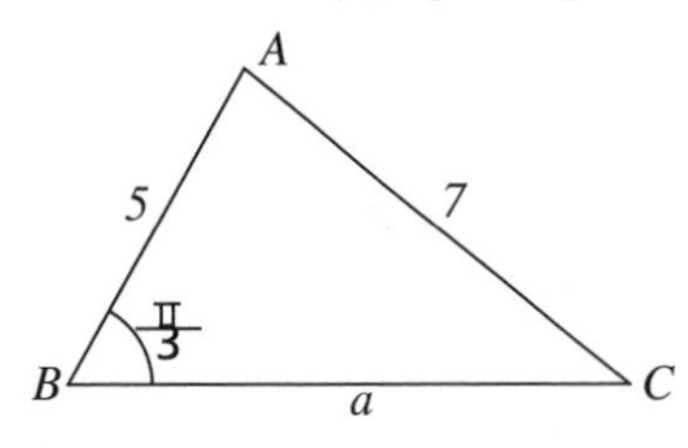

Figure 120

(a) Let $\overline{BC} = a$. Then, by the law of cosines,

$$a^2 + 5^2 - 2 \cdot 5 \cdot a \cdot \cos\frac{\pi}{3} = 7^2; \quad \text{i.e.,} \quad a^2 - 5a - 24 = 0.$$
$$(a-8)(a+3) = 0. \quad \therefore\ a = 8 \ \text{or} \ -3.$$

Because $a > 0$, we have $a = 8\ cm$.

Question: Can any meaning be given to the answer $a = -3$?

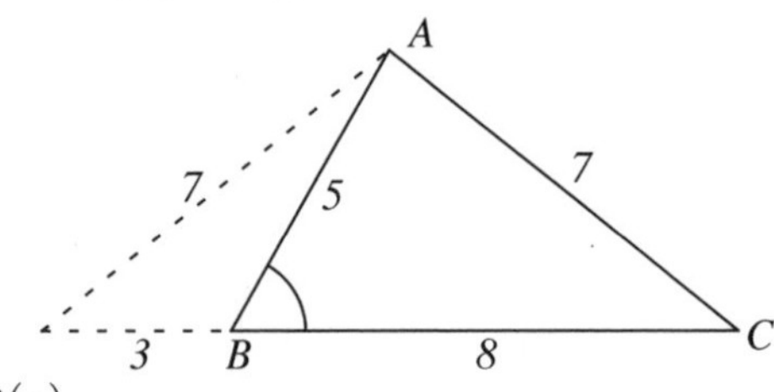

Figure 120(a)

(b)

$$[ABC] = \frac{1}{2}\overline{BC} \cdot \overline{AB} \sin(\angle ABC)$$
$$= \frac{1}{2} \cdot 8 \cdot 5 \cdot \frac{\sqrt{3}}{2} = 10\sqrt{3}\ cm^2.$$

Remark. Note that the known angle ($\angle ABC$) is not between two known sides (AB and AC). Yet the triangle is completely determined. (Why?)

Question: What if we interchange the lengths of AB and AC?

Exercise. Express the area of a triangle in terms of two of its angles and one side.

121. $[ABC] = 294\,cm^2$.

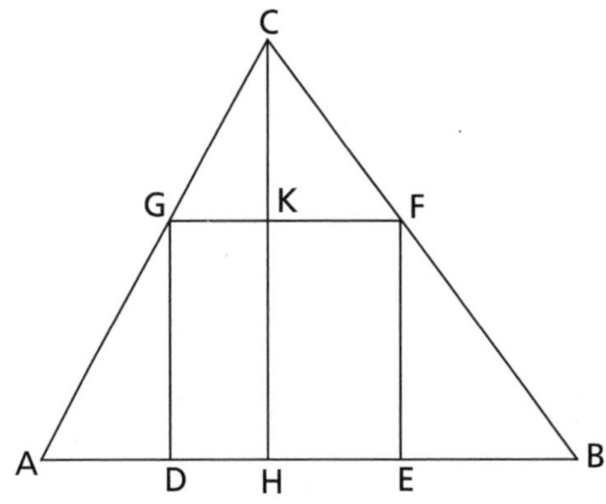

Figure 121(a)

Let K be the intersection of the altitude CH and FG. Because $\triangle ABC \sim \triangle GFC$, we have

$$\begin{aligned}
\frac{\overline{GF}}{\overline{AB}} &= \frac{\overline{CK}}{\overline{CH}} = \frac{\overline{CK}}{\overline{CK}+\overline{KH}}. \quad \therefore \frac{12}{28} = \frac{\overline{CK}}{\overline{CK}+12}, \\
\overline{CK} &= \frac{12^2}{28-12} = \frac{12^2}{16} = 9\,(cm). \\
\therefore \overline{CH} &= \overline{CK}+\overline{KH} = 9+12 = 21\,(cm). \\
\therefore [ABC] &= \frac{1}{2}\overline{AB}\cdot\overline{CH} = \frac{1}{2}\cdot 28\cdot 21 = 294\,(cm^2).
\end{aligned}$$

Remark. Note that our argument works even if either D or E is on the extension of AB.

122. (a) $R = \frac{5}{2}$. (b) $r = 1$.

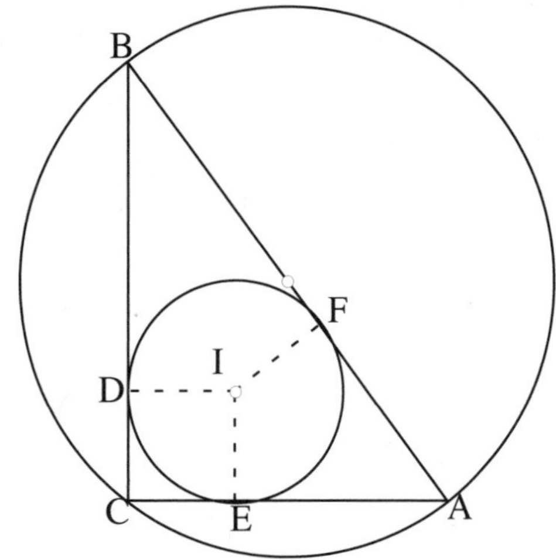

Figure 122(a)

(a) Clearly, the length of the hypotenuse is $2R$. $\therefore R = \frac{5}{2}$.

(b) Let D, E, F be the feet of the perpendiculars from incenter I to sides BC, CA, AB, respectively, of right triangle ABC (with $\angle C = \frac{\pi}{2}$). Then, $CEID$ is a square. Set $\overline{AE} = \overline{AF} = x$, $\overline{BD} = \overline{BF} = y$. Then

$$\begin{aligned}
r + x &= \overline{CE} + \overline{EA} = \overline{CA} = 3, \\
r + y &= \overline{CD} + \overline{DB} = \overline{BC} = 4, \\
x + y &= \overline{AF} + \overline{FB} = \overline{AB} = 5.
\end{aligned}$$

Solving this system of simultaneous equations, we obtain $r = 1$. ($x = 2$, $y = 3$.)

Exercise. Show that the inradius of an arbitrary Pythagorean triangle is an integer.

123. (a) $[ABC] = r(2R + r)$. (b) $2R$, $(R + r) \pm \sqrt{R^2 - 2Rr - r^2}$,
where R and r are the circumradius and the inradius, respectively.

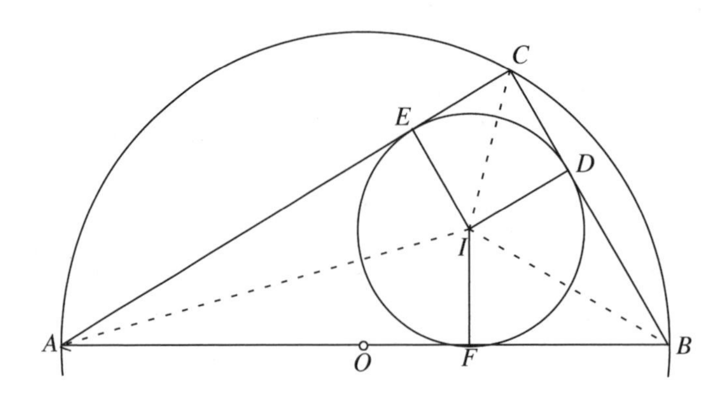

Figure 123

(a) Let D, E, F be the feet of the perpendiculars from the incenter I to the sides BC, CA, AB, respectively. Then

$$\begin{aligned}
\overline{ID} &= \overline{IE} = \overline{IF} = r. \\
\therefore\ [ABC] &= [IBC] + [ICA] + [IAB] \\
&= \frac{1}{2}\{\overline{ID}\cdot\overline{BC} + \overline{IE}\cdot\overline{CA} + \overline{IF}\cdot\overline{AB}\} \\
&= \frac{r}{2}\{(\overline{BD} + \overline{DC}) + (\overline{CE} + \overline{EA}) + \overline{AB}\} \\
&= \frac{r}{2}\{(\overline{BF} + r) + (r + \overline{AF}) + \overline{AB}\} \\
&= \frac{r}{2}\{2\overline{AB} + 2r\} = r(2R + r).
\end{aligned}$$

(b) It is clear that the length of the hypotenuse is $2R$. Let a, b be the lengths of the two legs. Then

$$a + b = 2(R + r), \quad ab = 2[ABC] = 2r(2R + r).$$

Therefore, a and b are the roots of the quadratic equation

$$\begin{aligned}
t^2 &- 2(R + r)t + 2r(2R + r) = 0. \\
\therefore\ t &= (R + r) \pm \sqrt{(R + r)^2 - 2r(2R + r)} \\
&= (R + r) \pm \sqrt{R^2 - 2Rr - r^2}.
\end{aligned}$$

124. There are 10 such points.

A picture is worth a thousand words.

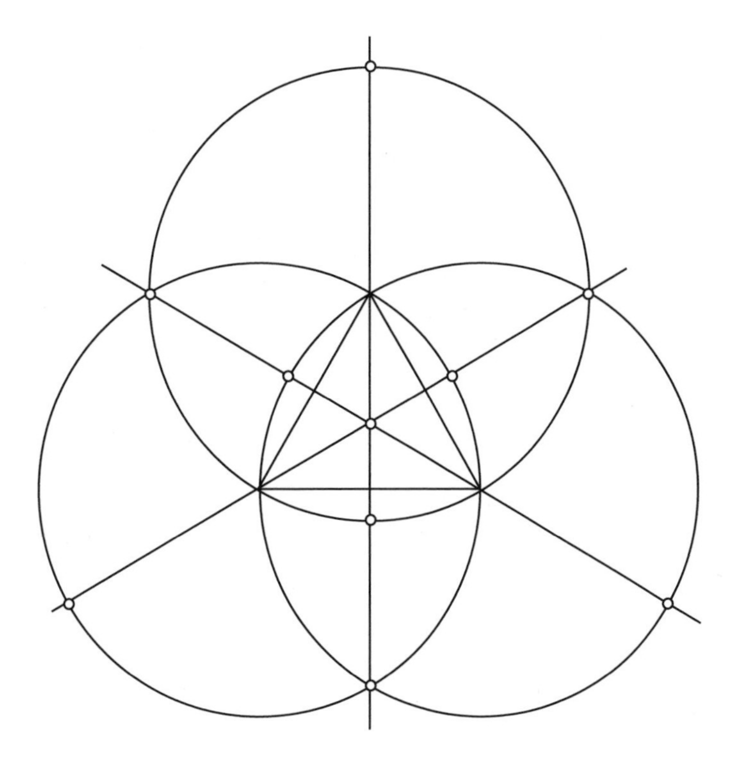

Figure 124

Exercise.

(a) Let P be a point in the plane of a square $ABCD$ such that triangles PAB, PBC, PCD, PDA are all isosceles. How many such points P are there?

(b) Let P be a point in the plane of a scalene triangle ABC such that at least two of

$$\triangle PBC, \quad \triangle PCA, \quad \triangle PAB$$

are isosceles. How many such points are there?

125. (a) A convex polygon can have, at most, three acute angles. (b) 8 sides.

(a) The exterior angle of an acute angle is obtuse, so if there were four (or more) acute angles, then the sum of all the exterior angles would exceed 2π. But the sum of all the exterior angles of any convex polygon is 2π. Therefore, there can be, at most, three acute interior angles, and in fact, any acute triangle is convex and has exactly three acute angles.

Remark. The reasoning above also shows that for each presence of a right angle, the number of acute angles reduces by one.

(b) By Part (a) and the remark, a convex polygon can have, at most, three acute angles, so if it has five obtuse angles, then there can be, at most, eight angles, and, in fact, there do exist convex octagons with five obtuse angles.

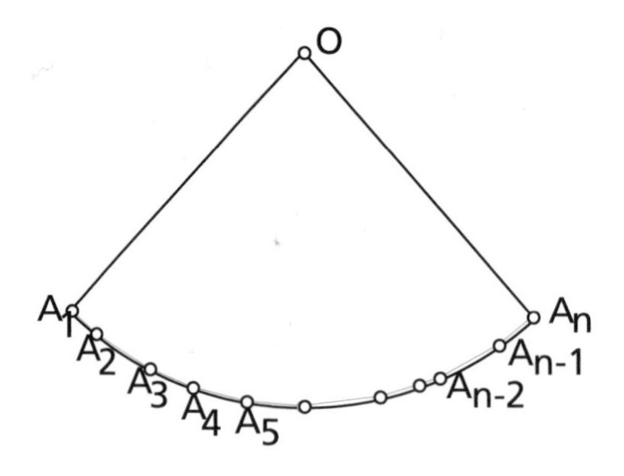

Figure 125

In Figure 125, choose n ($n \geq 2$) arbitrary points A_1, A_2, $\cdots$, A_n (in this order) on a circular arc, so that $\angle A_1OA_n$ is acute, where O is the center, and join the points O, A_1, A_2, $\cdots$, A_n, O successively, obtaining a polygon with $(n+1)$ sides. The three angles at O, A_1, and A_n are acute, and the remaining $(n-2)$ angles are obtuse. (In this case, $n = 7$.)

126. (a) 56. (b) 24.

(a) $\binom{8}{3} = \frac{8 \times 7 \times 6}{3 \times 2 \times 1} = 56.$

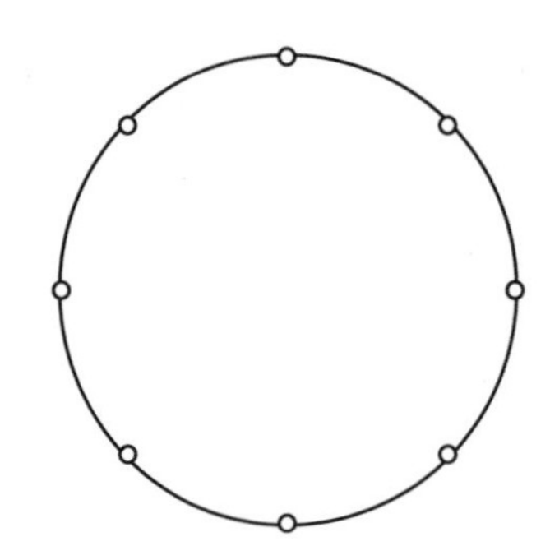

Figure 126

(b) The only way to obtain a right triangle is for the hypotenuse to be a diameter of the circle, and there are four possible diameters with two of given 8 points as ends. For each of these hypotenuses, there are six choices of the third vertex (corresponding to the right-angled vertex). Hence there are $4 \times 6 = 24$ right triangles with the eight given points as their vertices.

127. (a) 76. (b) 24.

(a) It is easy to see that there are four equilateral triangles. Consider the 12 points on the circle to be labelled 1, 2, $\cdots$, 12 just as on a clock. Then an angle is 60^0 if and only if it is subtended by an arc with two ends labelled m and n satisfying $m-n \equiv \pm 4$ (mod 12). There are 12 such arcs. For each of these 12 arcs, there are seven choices of the third vertex where the angle is 60^0. Of these seven choices, one of them gives an equilateral triangle which we have already counted. So for each of the 12 arcs, there are six choices of the third vertex which gives a triangle with exactly one 60^0 angle. Therefore, there are $12 \times 6 + 4 = 72 + 4 = 76$ triangles which have at least one 60^0 angle.

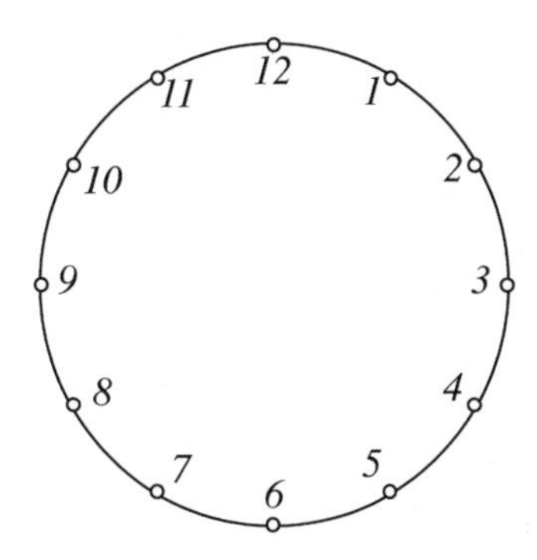

Figure 127

(b) An angle is 90^0 if and only if it is subtended by a diameter. For each of the six diameters, there are four choices of the third vertex that gives a triangle of the type 30^0-60^0-90^0. Hence there are $6 \times 4 = 24$ such triangles.

Exercise. Suppose there are 9 points on a circle, equally spaced. Of all the triangles having all their vertices at three of these 9 points, how many are acute triangles?

128. $n = 11$.

Of all the triangles having all their vertices at three of the vertices of a given convex n-gon, $n(n-4)$ of them share one side with the given convex n-gon (note that $n \geq 6$. Why?), and n of them share two sides. Therefore,

$$\binom{n}{3} - n(n-4) - n = 7n; \text{ i.e., } \frac{n(n-1)(n-2)}{1 \cdot 2 \cdot 3} - n(n+4) = 0,$$

which simplifies to

$$n(n+2)(n-11) = 0.$$

The only meaningful answer is $n = 11$.

129. (a) $\sqrt{2}$; 8 slices. (b) $\frac{1}{\sqrt{2}}$; 4 slices.

(a) At each vertex of a cube, there are three edges meeting at the vertex. A plane passing through the other three ends of the edges intersects the cube in an equilateral triangle whose sides are each $\sqrt{2}$. Clearly, there are eight such planes (one for each vertex).

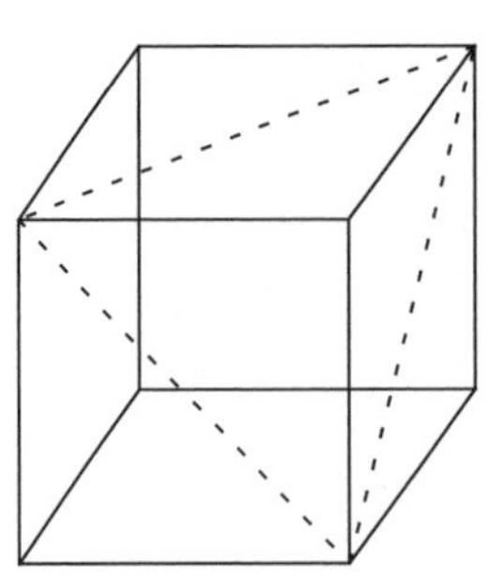
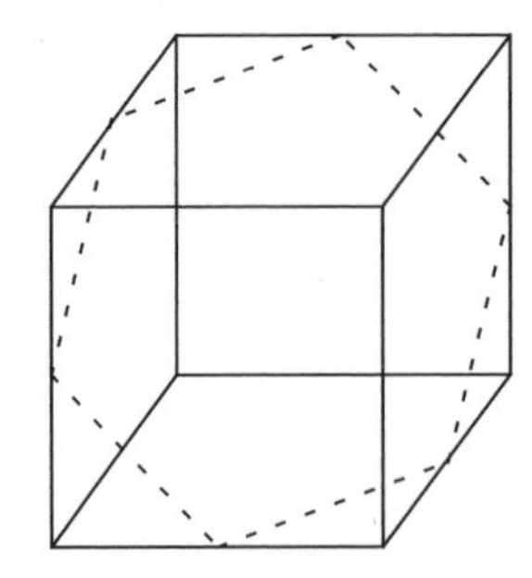

Figure 129

(b) For the cross-section to be a hexagon, the plane must intersect all six faces of the cube. And for the hexagon to be regular, all six sides must be of the same length, so the plane must meet every face 'in the same manner'. Thus the plane must meet the edges at their midpoints. We find that a plane that bisects a main diagonal (of length $\sqrt{3}$) perpendicularly produces the cross-section which is a regular hexagon whose sides have length $\frac{1}{\sqrt{2}}$. There are four such planes (one for each main diagonal).

Remark. Each of the planes in (b) lies exactly between a pair of parallel planes in (a).

Question: What if we try to slice the cube such that edges are cut into 1:2 ratio?

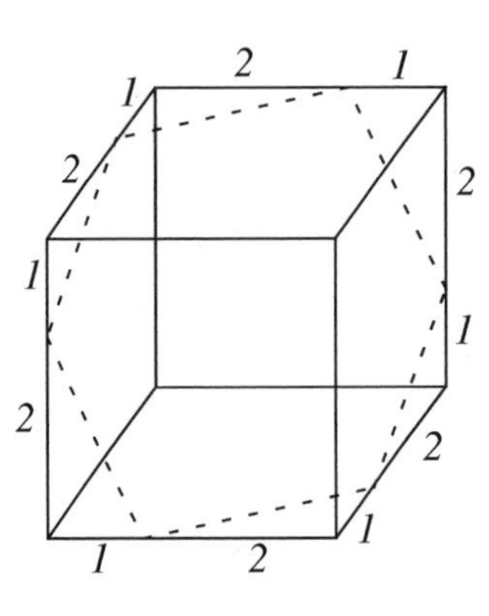

Figure 129(a)

130. (a) 5 *cm*; 2 ways. (b) 3 *cm*; infinitely many ways.

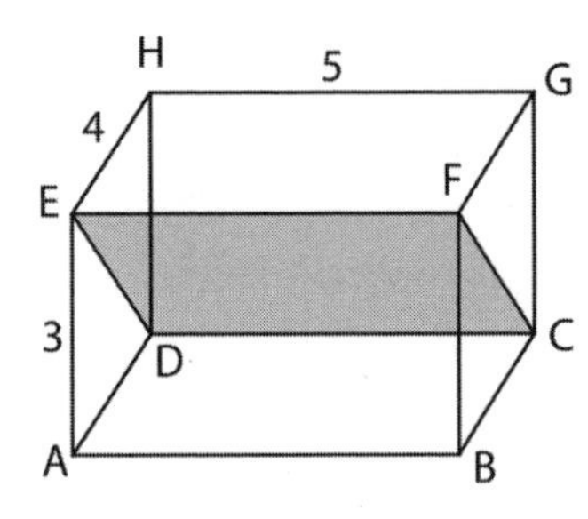

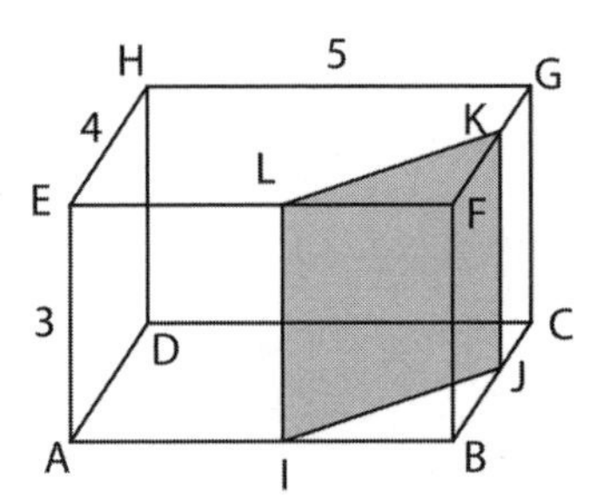

Figure 130(a)

(a) If the plane contains the edges AB and GH (or CD and EF), then the cross-section is a square whose sides have length 5 cm.

(b) There are infinitely many ways to slice the box by a plane so that the cross-section is a square whose sides have length 3 cm. For example, in Figure 130(a) (on the right), let

$$\overline{BI} = \overline{FL} = \sqrt{5}\ cm, \quad \overline{BJ} = \overline{FK} = 2\ cm;$$

then $IJKL$ is a square with the sides of length 3.

131. (a) 48. (b) The remaining eight triangles are equilateral with sides of length $\sqrt{2}$.

(We assume the sides of the cube are of unit length.)

(a) There are two types of right triangles:

For each diagonal of length $\sqrt{2}$ on a face of the cube there are two right triangles with side lengths 1, 1, $\sqrt{2}$. Because there are six faces and each face has two such diagonals, we have $2 \times 6 \times 2 = 24$ right triangles of this type.

For each edge there are two right triangles of side lengths 1, $\sqrt{2}$, $\sqrt{3}$. Because there are 12 edges, we have $2 \times 12 = 24$ triangles of this type.

Hence altogether we have $24 + 24 = 48$ right triangles.

(b) At each vertex there are three edges meeting there, and the other ends of these edges are the vertices of an equilateral triangle with sides of length $\sqrt{2}$. Because there are eight vertices, we have eight such equilateral triangles.

Because $\binom{8}{3} = \dfrac{8 \cdot 7 \cdot 6}{3 \cdot 2 \cdot 1} = 56$ and $48 + 8 = 56$, all the triangles have been accounted for.

132. The other two sides are either CE and EG or CM and IG.

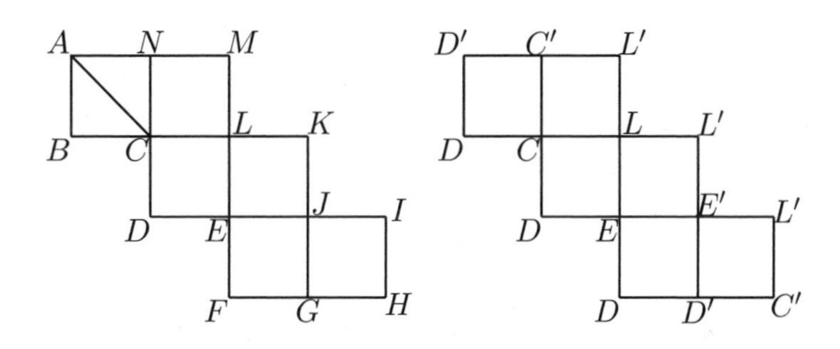

Figure 132(a)

Let us fold the figure on the left using the square $CDEL$ as a base and re-label the vertices above C, D, E, L as C', D', E', L', respectively. Then we obtain new labels as in the figure on the right.

By the discussion in the preceding problem, the vertices of an equilateral triangle are the other ends of three edges that meet at a vertex of a cube. The side AC is re-labelled as $D'C$, and D' and C can be regarded as

(a) the other ends of the edges DD' and DC that meet at D; or

(b) the other ends of the edges $C'D'$ and $C'C$ that meet at C'.

In case (a), the three edges that meet at D are DD', DC, and DE. Thus $D'C$, CE, ED' (in the figure on the right) are the three sides of an equilateral triangle. In terms of the original figure on the left, they are AC, CE, and EG.

In case (b), the three edges that meet at C' are $C'D'$, $C'C$, and $C'L'$. Thus $D'C$, CL', $L'D'$ (in the figure on the right) are the three sides of an equilateral triangle. In terms of the original figure on the left, they are AC, CM, and IG.

133. (a) 72 ways. (b) 420 ways.

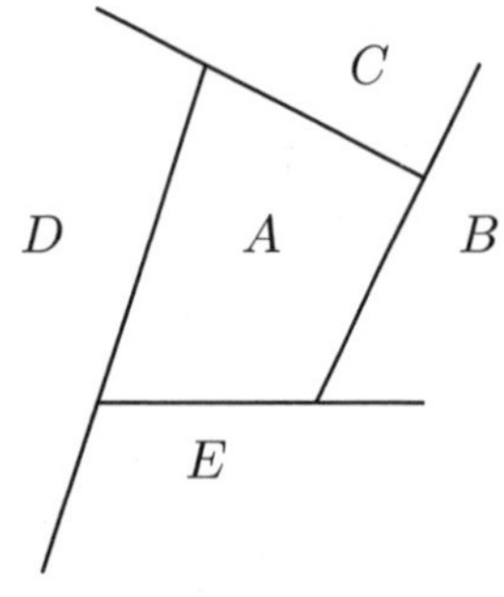

Figure 133

(a) There are $4 \cdot 3 \cdot 2$ ways to color the regions A, B, C.

(i) If we color the region D with the same color as for the region B, then we have two choices of color for the region E.

(ii) If we color the region D with a color different from that for B (i.e., using the remaining color), then we must color the region E with the color of the region C.

Therefore, there are $4 \cdot 3 \cdot 2(1 \cdot 2 + 1 \cdot 1) = 2^3 \cdot 3^2 = 72$ ways to color the regions.

(b) There are $5 \cdot 4 \cdot 3$ ways to color the regions A, B, C.

(i) If we color the region D with the same color as for the region B, then we have 3 choices of colors for the region E.

(ii) If we color the region D with a color different from that for B, then we have 2 choices each for the regions D and E.

Therefore, there are $5 \cdot 4 \cdot 3(1 \cdot 3 + 2^2) = 5 \cdot 4 \cdot 3 \cdot 7 = 420$ ways to color the regions.

134. (a) 30 ways. (b) 180 ways.

(a) Fix one color at the bottom. Then there are five choices for the top. The remaining four faces on the sides can be colored in 4! ways, except that for each way of coloring, we obtain, by rotation, 3 'other' colorings. So all together, we have $5 \cdot \frac{4!}{4} = 5 \cdot 3 \cdot 2 = 30$ ways to color a cube.

(b) This time we cannot fix a color at the bottom, because the faces have different sizes. There are $\binom{6}{2}$ ways to choose two colors for the pair of the biggest faces. (Note that once two colors are chosen, two ways of painting are the same.) Now there are $\binom{4}{2}$ ways to choose two colors for the pair of the smallest faces. (Again, once two colors are chosen, two ways of painting are the same—simply turn the box and look from the opposite end.) Finally, use the remaining two colors to paint the remaining two faces. This time two ways of painting are different! So altogether, we have $\binom{6}{2} \cdot \binom{4}{2} \cdot 2 = \frac{6!}{(2!)^2} = 180$ ways to color a rectangular box.

135. 17 *cm*.

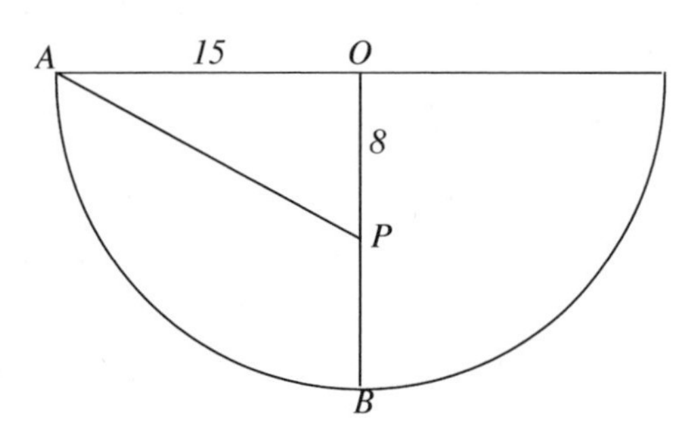

Figure 135(a)

Because $\overline{AB} = \overline{OA}$, if we cut along OA, then the cone becomes a semi-circle (with radius 15), and $OA \perp OB$. Therefore, by the Pythagorean theorem, $\overline{AP} = \sqrt{15^2 + 8^2} = \sqrt{289} = 17$ *cm*.

136. The figures which can be so folded are Figures (d), (f), (g), (h).

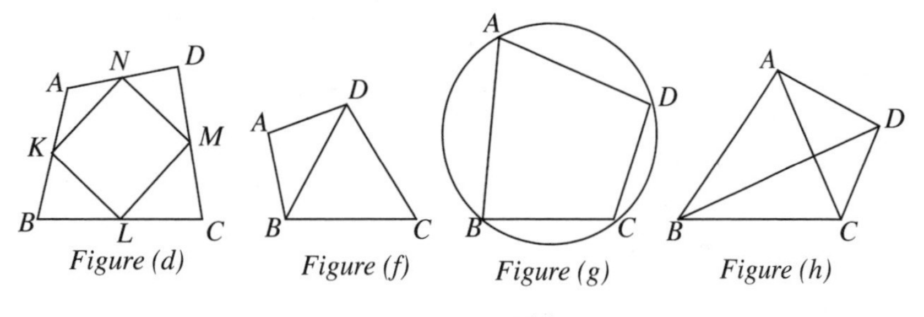

Figure 136(a)

Question: What is the necessary and sufficient condition?

137. The perimeter of $\triangle EBG$ is 2ℓ.

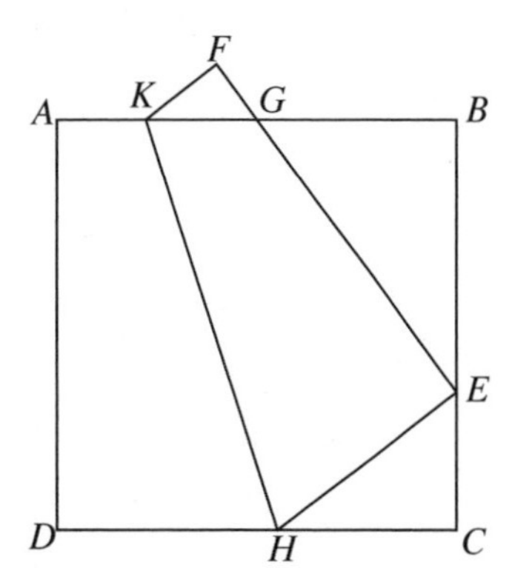

Figure 137

Before we prove that the perimeter is of fixed length, we should find out what this fixed length should be (assuming the problem is correct).

So we do some experiments. If E is very close to C, then $\triangle EBG$ will be very narrow and almost coincide with the side BC. Therefore, in this case, the perimeter of $\triangle EBG$ will be almost $2\overline{BC} = 2\ell$.

If E is very close to B, then $\triangle EBG$ will again be very narrow and almost coincide with the side AB. So, again in this case, the perimeter of $\triangle EBG$ will be almost $2\overline{AB} = 2\ell$.

Therefore, we see that the fixed perimeter (assuming that the perimeter really is constant) must be 2ℓ.

Now we set out to prove our conjecture. Let $\overline{BE} = a$, $\overline{BG} = b$, $\overline{EG} = c$. Because $\triangle HCE \sim \triangle EBG$, we have

$$\frac{\overline{CH}}{a} = \frac{\overline{HE}}{c} = \frac{\overline{CE}}{b} = \frac{\ell - a}{b}. \quad \therefore \ \ell = \overline{CH} + \overline{HE} = \frac{\ell - a}{b}(a + c).$$

Solving for ℓ, we obtain

$$\ell = \frac{a(a+c)}{a-b+c}.$$

If our conjecture is correct, we must have

$$2\frac{a(a+c)}{a-b+c} = a + b + c.$$

Clearing the denominator and using the Pythagorean theorem: $a^2 + b^2 = c^2$, we easily verify this last equality, so we are done.

Remark. The last part of the computation can also be carried out as follows[4]:

$$\begin{aligned} \ell &= \frac{a(a+c)}{a-b+c} = \frac{a(a+c)}{a-b+c} \cdot \frac{a+b+c}{a+b+c} \\ &= \frac{a(a+c)(a+b+c)}{a^2+2ac+c^2-b^2} = \frac{a(a+c)(a+b+c)}{2a(a+c)} \qquad (\because\ a^2+b^2=c^2) \\ &= \frac{1}{2}(a+b+c). \end{aligned}$$

Alternate Solution. Let $\overline{EB} = a$, $\overline{BG} = b$, $\overline{EG} = c$, as before. Because $\triangle HCE$, $\triangle KFG$, $\triangle EBG$ are similar, let the ratio of the lengths of their corresponding sides be $r_1 : r_2 : 1$. Then

$$\begin{aligned} \ell &= \overline{BE} + \overline{CE} = a + r_1 b, \\ \ell &= \overline{EG} + \overline{FG} = c + r_2 b, \\ \ell &= \overline{CH} + \overline{EH} = r_1(a+c), \\ \ell &= \overline{BG} + \overline{GK} + \overline{KF} = b + r_2(c+a). \end{aligned}$$

[4] One of the earliest mathematics tricks we learn in elementary school is to express 1 in a fancy way. For example,

$$\frac{2}{3} + \frac{4}{5} = \frac{2}{3} \cdot \frac{5}{5} + \frac{4}{5} \cdot \frac{3}{3} = \frac{2 \cdot 5 + 4 \cdot 3}{3 \cdot 5} = \frac{22}{15}.$$

Adding the first two equalities, we obtain

$$2\ell = a + c + (r_1 + r_2)b.$$

Adding the third and the fourth equalities, we obtain

$$2\ell = (r_1 + r_2)(a + c) + b.$$

Subtracting the last two equalities, we obtain

$$(a + c)(1 - r_1 - r_2) + (r_1 + r_2 - 1)b = 0.$$

$$\therefore\ (a - b + c)(1 - r_1 - r_2) = 0.$$

But $a + c > b$, and so $r_1 + r_2 = 1$.

$$\therefore\ a + b + c = 2\ell.$$

Remark. The sum of the perimeters of the three similar triangles is equal to the perimeter of the square.

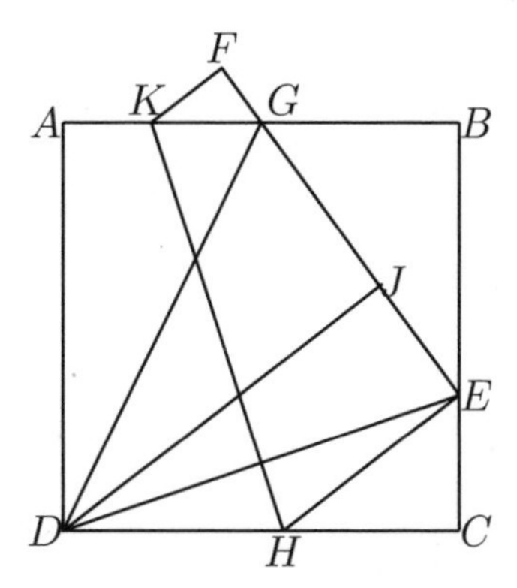

Figure 137(a)

Geometric Solution. Let J be the foot of the perpendicular from D to EF. Then, because HK is the perpendicular bisector of DE,

$$\angle HDE = \angle HED = \angle EDJ \quad (\because\ DJ \parallel HE).$$

Now, right triangles DEJ and DEC share the hypotenuse DE and two pairs of corresponding angles are equal.

$$\therefore\ \triangle DEJ \cong \triangle DEC.$$

It follows that $\overline{EJ} = \overline{EC}$, and $\overline{DJ} = \overline{DC}$. Now, $\triangle DGJ$ and $\triangle DGA$ are right triangles with two pairs of corresponding sides equal.

$$\therefore\ \triangle DGJ \cong \triangle DGA, \quad \text{and so} \quad \overline{JG} = \overline{AG}.$$

Hence

$$\begin{aligned} \overline{BE} + \overline{EG} + \overline{GB} &= \overline{BE} + (\overline{EJ} + \overline{JG}) + \overline{GB} \\ &= (\overline{BE} + \overline{EC}) + (\overline{AG} + \overline{GB}) \\ &= \overline{BC} + \overline{AB} = 2\ell. \end{aligned}$$

Experiment. Take a 6×6 square sheet of paper and choose E to be the midpoint of the side BC; determine the length of each side of $\triangle EBG$.

138. $[EFGH] = 126\ cm^2$.

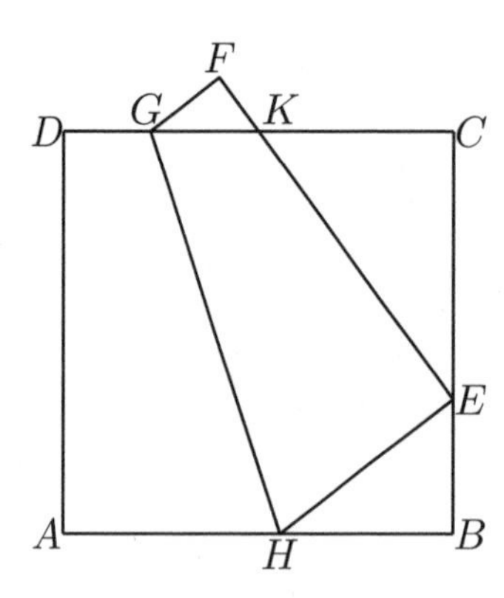

Figure 138

Let $\overline{BH} = x$. Then

$$\overline{EH} = \overline{AH} = \overline{AB} - \overline{BH} = 18 - x.$$

By the Pythagorean theorem, we have

$$(18 - x)^2 = x^2 + 6^2.$$

Solving for x, we obtain

$$2 \cdot 18x = 18^2 - 6^2 = 24 \cdot 12.$$

$$\therefore\ \overline{BH} = x = \frac{24 \cdot 12}{2 \cdot 18} = 8. \quad \overline{EH} = 18 - x = 18 - 8 = 10.$$

Because $\triangle CEK \sim \triangle BHE$ (where K is the intersection of CD and EF), we have

$$\frac{\overline{EK}}{\overline{HE}} = \frac{\overline{CK}}{\overline{BE}} = \frac{\overline{CE}}{\overline{BH}}; \quad \frac{\overline{EK}}{10} = \frac{\overline{CK}}{6} = \frac{18 - 6}{8}.$$

$$\therefore\ \overline{EK} = \frac{12 \cdot 10}{8} = 15, \quad \overline{CK} = \frac{12 \cdot 6}{8} = 9.$$

Because $\triangle FGK \sim \triangle CEK$, we have

$$\frac{\overline{FG}}{\overline{FK}} = \frac{\overline{CE}}{\overline{CK}}. \quad \therefore\ \overline{FG} = \frac{\overline{CE}}{\overline{CK}} \cdot \overline{FK} = \frac{12}{9} \cdot (18 - 15) = 4.$$

$$\therefore\ [EFGH] = \frac{1}{2}(\overline{EH} + \overline{FG}) \cdot \overline{EF} = \frac{1}{2}(10 + 4) \cdot 18 = 126\ cm^2.$$

Remark. Once we know that $\overline{FG} = 4$, we obtain

$$\overline{CG} = \overline{CD} - \overline{DG} = \overline{CD} - \overline{FG} = 18 - 4 = 14.$$

Or, knowing that $\triangle FGK$ is a right triangle with $\overline{FK} = 3$, $\overline{FG} = 4$, we have $\overline{GK} = 5$.

$$\therefore\ \overline{CG} = \overline{CK} + \overline{GK} = 9 + 5 = 14.$$

Hence

$$\begin{aligned}[EFGH] &= [ADGH] = [ABCD] - [BCGH] \\ &= 18^2 - \frac{1}{2}\left(\overline{BH} + \overline{CG}\right) \cdot \overline{BC} \\ &= 18^2 - \frac{1}{2}(8+14) \cdot 18 \\ &= 18 \cdot (18 - 11) = 18 \cdot 7 = 126\ cm^2.\end{aligned}$$

139. The length of a side is $\sqrt{29}\ cm$.

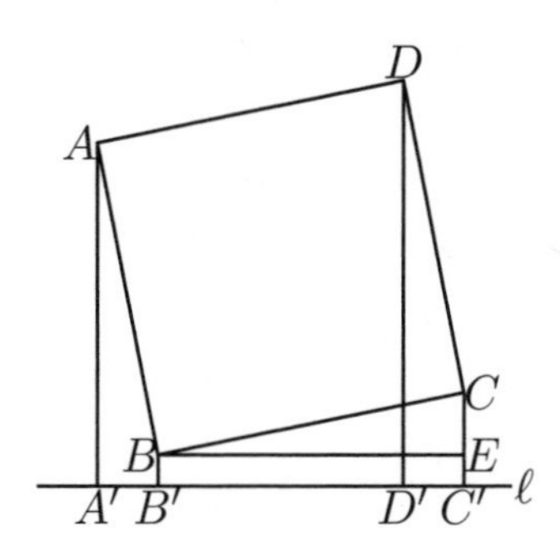

Figure 139(a)

$$\begin{aligned}\overline{BC} &= \sqrt{\overline{BE}^2 + \overline{CE}^2} \\ &= \sqrt{\overline{B'C'}^2 + \overline{C'D'}^2} \\ &= \sqrt{(3+2)^2 + 2^2} = \sqrt{29}.\end{aligned}$$

140. (a) $[ABCD] = 12\ cm^2$. (b) $\overline{AB} = \sqrt{2}\ cm$, $\overline{BC} = 6\sqrt{2}\ cm$.

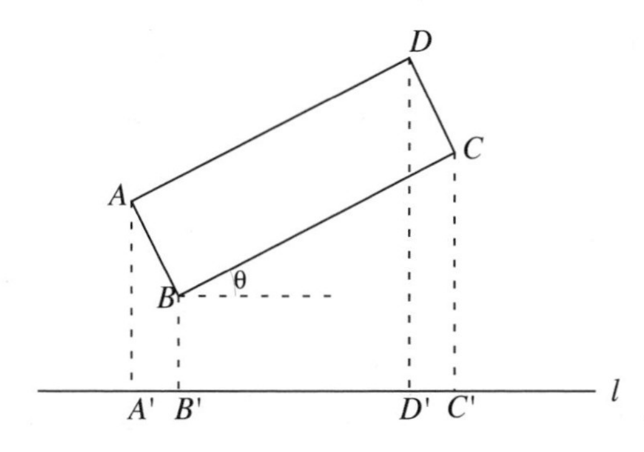

Figure 140

Let θ be the angle between BC and ℓ. Then

$$\begin{aligned}\overline{BC} \cdot \cos\theta &= \overline{B'C'} = \overline{B'D'} + \overline{D'C'} = 5 + 1 = 6. \\ \overline{AB} \cdot \sin\theta &= \overline{A'B'} = 1. \\ \therefore\ [ABCD] &= \overline{AB} \cdot \overline{BC} = \frac{1}{\sin\theta} \cdot \frac{6}{\cos\theta} = \frac{6}{\frac{1}{2} \cdot \sin 2\theta} = \frac{12}{\sin 2\theta}.\end{aligned}$$

Thus, $[ABCD]$ is minimized when $\sin 2\theta = 1$; i.e., when $\theta = \dfrac{\pi}{4}$. In this case, $[ABCD] = 12\ cm^2$, and

$$\overline{AB} = \frac{1}{\sin\frac{\pi}{4}} = \sqrt{2}\ cm, \quad \overline{BC} = \frac{6}{\cos\frac{\pi}{4}} = 6\sqrt{2}\ cm.$$

141. The proposition is correct. Observe that the areas of $\triangle AJD$ and $\triangle CFB$ are 1/3 of that of $\triangle ACD$ and $\triangle ABC$, respectively. Hence their sum is 1/3 of that of quadrangle $ABCD$. Thus, the area of quadrangle $AFCJ$ is 2/3 of that of $ABCD$. Now, the area of quadrangle $EFIJ$, being the sum of that of $\triangle EFJ$ and $\triangle FIJ$, is clearly 1/2 of that of quadrangle $AFCJ$, and so is 1/3 of that of quadrangle $ABCD$.

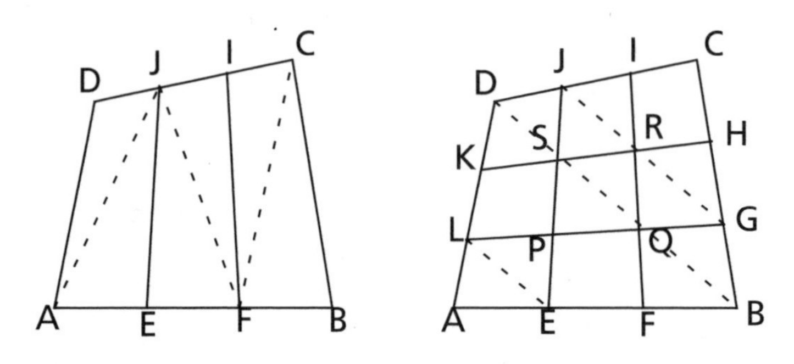

Figure 141(a)

Hence if we can show that P, S are trisection points of EJ, and Q, R are trisection points of FI, then by precisely the same argument, we obtain

$$[PQRS] = \frac{1}{3}[EFIJ] = \frac{1}{3}\cdot\frac{1}{3}[ABCD] = \frac{1}{9}[ABCD],$$

and our proof will be complete. Now

$$\overline{AE} = \frac{1}{3}\overline{AB}, \quad \overline{AL} = \frac{1}{3}\overline{AD}.$$
$$\therefore EL \parallel BD, \quad \overline{EL} = \frac{1}{3}\overline{BD}.$$

Similarly,

$$\overline{CG} = \frac{2}{3}\overline{CB}, \quad \overline{CJ} = \frac{2}{3}\overline{CD}.$$
$$\therefore GJ \parallel BD, \quad \overline{GJ} = \frac{2}{3}\overline{BD}.$$

It follows that $EL \parallel GJ$ and $\overline{EL} = \frac{1}{2}\overline{JG}$. Therefore, $\triangle PLE \sim \triangle PGJ$ and

$$\frac{\overline{PE}}{\overline{PJ}} = \frac{\overline{LE}}{\overline{GJ}} = \frac{1}{2}. \quad \therefore \overline{EP} = \frac{1}{2}\overline{JP} = \frac{1}{3}\overline{EJ}.$$

Similarly,

$$\overline{JS} = \frac{1}{3}\overline{JE}, \quad \text{and} \quad \overline{FQ} = \frac{1}{3}\overline{FI} = \overline{IR}.$$

Therefore, P, S are trisection points of EJ, and Q, R are trisection points of FI, and so our proof is complete.

142. The area of the parallelogram is $5\sqrt{3}\ cm^2$.

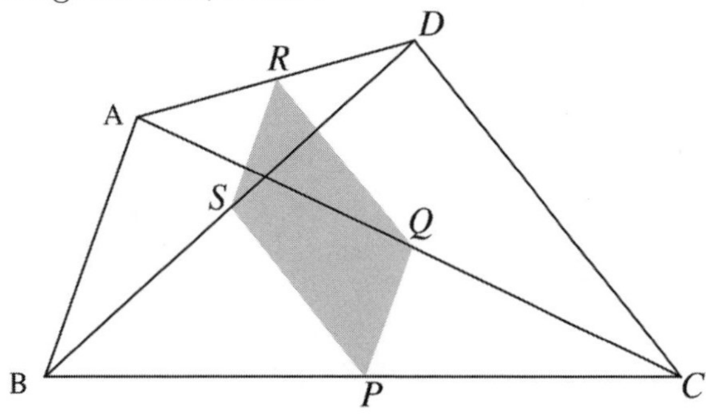

Figure 142

Because P and Q are the midpoints of two sides of $\triangle ABC$, we have

$$\begin{aligned} \overline{PQ} &= \frac{1}{2}\overline{AB} = \frac{5}{2}, \qquad PQ \parallel AB. \\ \therefore\ \angle QPC &= \angle ABC = 70^0. \end{aligned}$$

Similarly,

$$\overline{RS} = \frac{1}{2}\overline{AB}, \qquad RS \parallel AB \parallel PQ.$$

Hence $PQRS$ is a parallelogram. In a similar manner,

$$\begin{aligned} \overline{PS} &= \frac{1}{2}\overline{CD} = 4, \qquad PS \parallel CD. \\ \therefore\ \angle BPS &= \angle BCD = 50^0. \end{aligned}$$

It follows that

$$\begin{aligned} \angle QPS &= 180^0 - (\angle QPC + \angle BPS) \\ &= 180^0 - (70^0 + 50^0) = 60^0. \\ \therefore [PQRS] &= 2\,[PQS] = 2 \cdot \frac{1}{2} \cdot \overline{PQ} \cdot \overline{PS} \cdot \sin(\angle QPS) \\ &= \frac{5}{2} \cdot 4 \cdot \frac{\sqrt{3}}{2} = 5\sqrt{3}\ cm^2. \end{aligned}$$

Question: Note that there is no condition on the length of the side BC. What if C approaches B along a straight line?

143. (a) Either $[ABC] = 5\ cm^2$ or $[ABC] = \frac{15}{2}\ cm^2$.

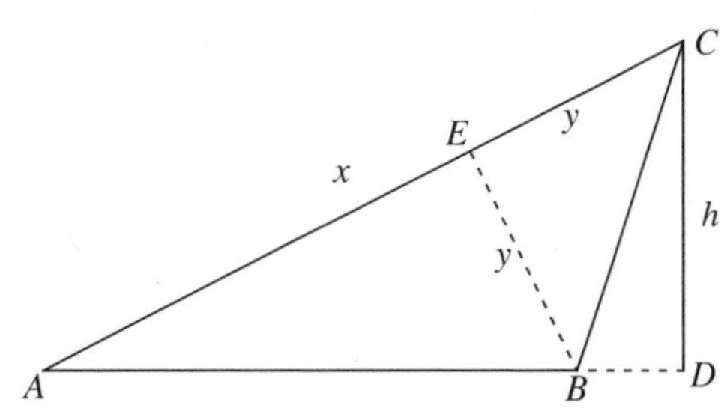

Figure 143(a)

(a) Let E be the foot of the perpendicular from the vertex B to the side CA, and

$$\overline{CA} = x, \quad \overline{CE} = \overline{BE} = y, \quad \overline{CD} = h.$$

Then $\overline{AE}^2 + \overline{BE}^2 = \overline{AB}^2$; i.e.,

$$(x-y)^2 + y^2 = 5^2; \text{ and so } x^2 - 2xy + 2y^2 = 25.$$

Also $\overline{CD}^2 = \overline{CA}^2 - \overline{AD}^2 = \overline{BC}^2 - \overline{BD}^2$ (where $\overline{BC} = \sqrt{2}y$); i.e.,

$$h^2 = x^2 - 6^2 = (\sqrt{2}y)^2 - 1^2, \text{ and so } x^2 - 2y^2 = 6^2 - 1^2 = 35.$$

$$\therefore\ 7(x^2 - 2xy + 2y^2) - 5(x^2 - 2y^2) = 7 \cdot 25 - 5 \cdot 35 = 0;$$

i.e.,

$$2x^2 - 14xy + 24y^2 = 2(x-3y)(x-4y) = 0.$$

(i) If $x = 3y$, then $7y^2 = 35$, $y = \sqrt{5}$, $x = 3\sqrt{5}$.

$$\therefore\ [ABC] = \frac{1}{2}xy = \frac{1}{2}\sqrt{5}\,(3\sqrt{5}) = \frac{15}{2}\ cm^2.$$

(ii) If $x = 4y$, then $14y^2 = 35$, $y = \sqrt{\frac{5}{2}}$, $x = 4\sqrt{\frac{5}{2}}$.

$$\therefore\ [ABC] = \frac{1}{2}xy = \frac{1}{2}\left(4\sqrt{\frac{5}{2}}\right)\sqrt{\frac{5}{2}} = 5\ cm^2.$$

A Variation of the Solution above. Computing the area of the triangle ABC in two ways, we obtain

$$\begin{aligned} \frac{1}{2}5h &= \frac{1}{2}xy, \\ \therefore 25h^2 &= x^2y^2 = (h^2+6^2)\left(\frac{h^2+1}{2}\right), \\ 50h^2 &= h^4 + 37h^2 + 36; \\ \therefore\ h^4 - 13h^2 + 36 &= (h^2-4)(h^2-9) = 0. \end{aligned}$$

Because $h > 0$, we obtain

$$h = 2,\ [ABC] = 5; \quad \text{or } h = 3,\ [ABC] = \frac{15}{2}.$$

Trigonometric Solution. Let $\overline{CD} = h$ as before. Because $\angle ACD - \angle BCD = \angle ACB = \frac{\pi}{4}$, we have

$$\begin{aligned} 1 &= \tan\frac{\pi}{4} = \tan(\angle ACD - \angle BCD) \\ &= \frac{\tan(\angle ACD) - \tan(\angle BCD)}{1 + \tan(\angle ACD)\cdot\tan(\angle BCD)} \\ &= \frac{\frac{6}{h} - \frac{1}{h}}{1 + \frac{6}{h}\cdot\frac{1}{h}} = \frac{5h}{h^2+6}. \\ \therefore\ h^2 - 5h + 6 &= (h-3)(h-2) = 0. \end{aligned}$$

(i) If $h = 3$, then $[ABC] = \frac{1}{2} \cdot \overline{AB} \cdot h = \frac{1}{2} \cdot 5 \cdot 3 = \frac{15}{2}\ cm^2$.

(ii) If $h = 2$, then $[ABC] = \frac{1}{2} \cdot \overline{AB} \cdot h = \frac{1}{2} \cdot 5 \cdot 2 = 5\ cm^2$.

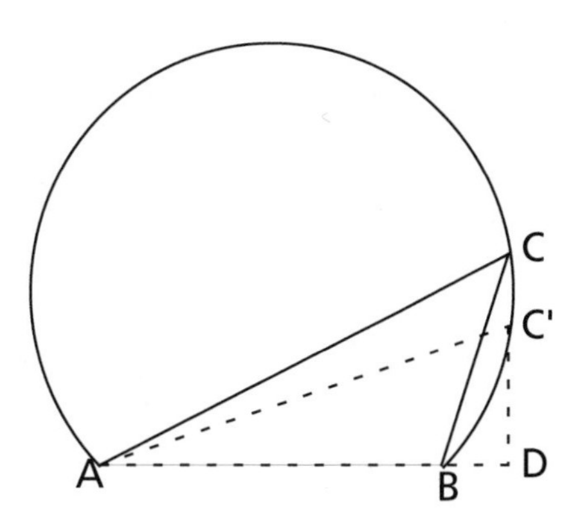

Figure 143(b)

(b) There are two solutions, because for fixed line segment AB, the locus of a point C such that $\angle ACB = \frac{\pi}{4}$ is a circular arc, to be precise (neglecting the mirror image with respect to the base AB), $\frac{3}{4}$ of a circle, which intersects the line perpendicular to AB at D at two points.

Exercise. In an acute triangle ABC, let D be the foot of the perpendicular from C to AB. Suppose $\angle ACB = \frac{\pi}{4}$ and $\overline{AD} = 3\ cm$, $\overline{BD} = 2\ cm$. Find the area of $\triangle ABC$.

144. $\overline{AD} = 10$, $\overline{AE} = 12$; or $\overline{AD} = 12$, $\overline{AE} = 10$.

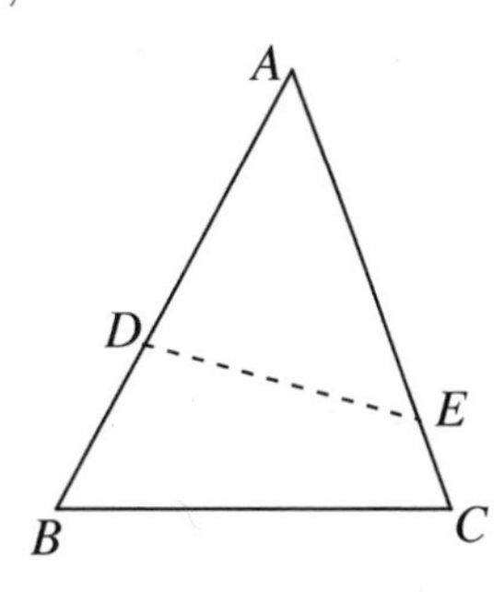

Figure 144

Let $\overline{AD} = x$, $\overline{AE} = y$. Then because $2[ADE] = [ABC]$, we have

$$\begin{aligned} 2 \cdot \frac{1}{2} \cdot \overline{AD} \cdot \overline{AE} \cdot \sin A &= \frac{1}{2} \cdot \overline{AB} \cdot \overline{AC} \cdot \sin A, \\ \therefore\ xy &= \frac{1}{2} \cdot 16 \cdot 15 = 120. \end{aligned}$$

Also,

$$\begin{aligned}\overline{AD} + \overline{AE} &= \frac{1}{2}\{\overline{AB} + \overline{BC} + \overline{CA}\};\\ \therefore\ x + y &= \frac{1}{2}(16 + 13 + 15) = 22.\end{aligned}$$

It follows that x and y are solutions of the quadratic equation

$$\begin{aligned}t^2 - 22t + 120 &= (t-10)(t-12) = 0.\\ \therefore\ (x,\ y) &= (10,\ 12) \text{ or } (12,\ 10).\end{aligned}$$

145. $[ABC] = 36\ cm^2$.

A picture is worth a thousand words.

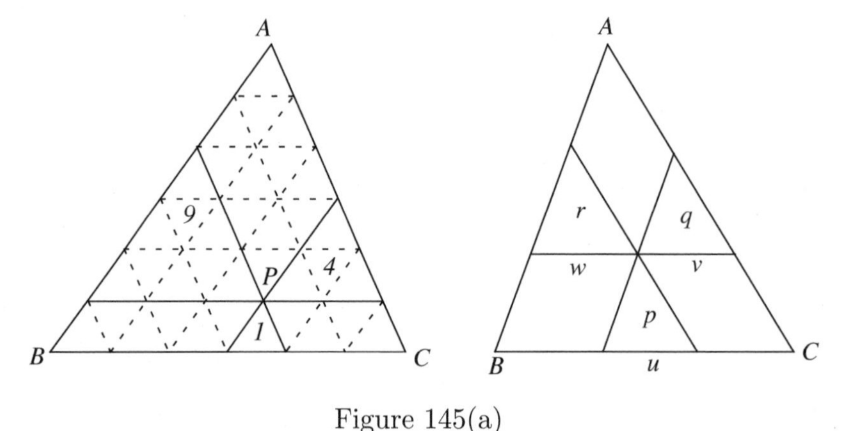

Figure 145(a)

General Solution. Note that all three small triangles are similar to the original triangle ABC. Let p, q, r be the areas of the three (small) triangles, and u, v, w the lengths of the sides corresponding to BC. Because the ratio of the areas of similar triangles is equal to that of the squares of the lengths of the corresponding sides, we have

$$\begin{aligned}\frac{[ABC]}{\overline{BC}^2} &= \frac{p}{u^2} = \frac{q}{v^2} = \frac{r}{w^2}\\ &= \frac{\sqrt{qr}}{vw} = \frac{\sqrt{rp}}{wu} = \frac{\sqrt{pq}}{uv}\\ &= \frac{(p+q+r) + 2(\sqrt{qr} + \sqrt{rp} + \sqrt{pq})}{(u^2+v^2+w^2) + 2(vw+wu+uv)}\\ &= \frac{\left(\sqrt{p} + \sqrt{q} + \sqrt{r}\right)^2}{(u+v+w)^2},\end{aligned}$$

where we have used the result in the exercise at the end of the solution to Problem 34. Because $u + v + w = \overline{BC}$, we obtain

$$[ABC] = \left(\sqrt{p} + \sqrt{q} + \sqrt{r}\right)^2.$$

In our case, we have

$$[ABC] = \left(\sqrt{1} + \sqrt{4} + \sqrt{9}\right)^2 = 36\ cm^2.$$

Question: What if point P moves outside $\triangle ABC$?

146. $[ABC] = 50\ cm^2$.

A picture is worth a thousand words.

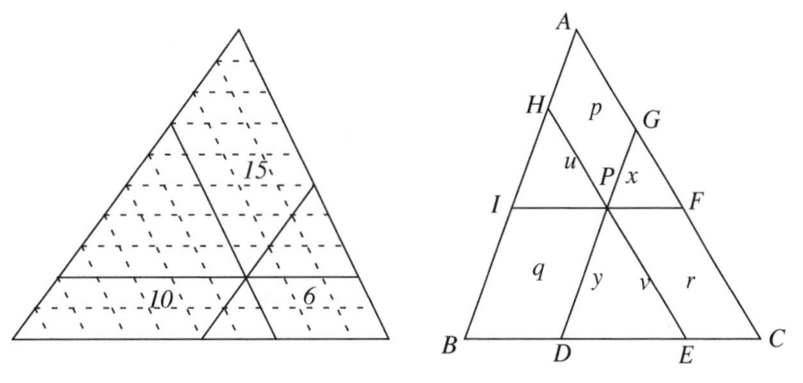

Figure 146(a)

General Solution. Let p, q, r be the areas of the three parallelograms, and x, y, u, v the lengths of sides as in Figure 146(a). Observe that the two parallelograms $AHPG$ and $BDPI$ have the same height, so the ratio of their areas is equal to that of the lengths of their base. Therefore $\dfrac{x}{p} = \dfrac{y}{q}$. Similarly, $\dfrac{u}{p} = \dfrac{v}{r}$. $\therefore$ $\dfrac{xu}{p^2} = \dfrac{yv}{qr}$.

$$\begin{aligned}[PDE] &= \frac{1}{2} yv \sin(\angle A) \\ &= \frac{1}{2}\left(\frac{qr}{p^2}\right) \cdot xu \sin(\angle A) = \frac{qr}{2p}.\end{aligned}$$

In a similar manner,

$$[PFG] = \frac{rp}{2q}, \quad [PHI] = \frac{pq}{2r}.$$

$$\begin{aligned}\therefore\ [ABC] &= [AHPG] + [BDPI] + [CFPE] \\ &\quad + [PDE] + [PFG] + [PHI] \\ &= p + q + r + \frac{1}{2}\left(\frac{qr}{p} + \frac{rp}{q} + \frac{pq}{r}\right).\end{aligned}$$

In our case, we have

$$\begin{aligned}[ABC] &= 6 + 10 + 15 + \frac{1}{2}\left(\frac{10 \cdot 15}{6} + \frac{15 \cdot 6}{10} + \frac{6 \cdot 10}{15}\right) \\ &= 31 + \frac{1}{2}\left(5^2 + 3^2 + 2^2\right) = 50\ cm^2.\end{aligned}$$

Question: What if point P moves outside $\triangle ABC$?

147. $[ABCD] = 121\ cm^2$.

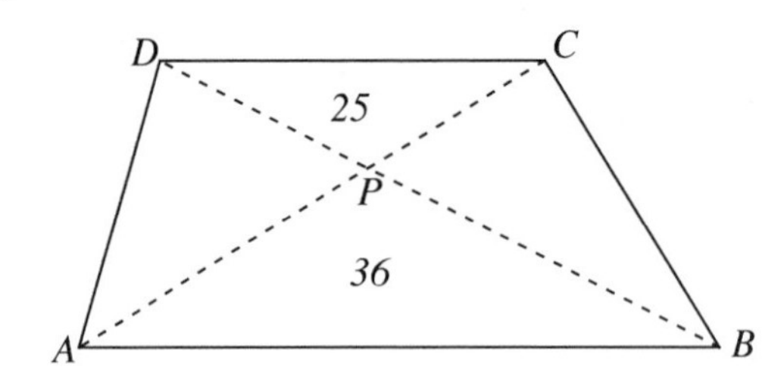

Figure 147

Let $\ell_1 = \overline{AB}$, $\ell_2 = \overline{CD}$, and the heights of $\triangle PAB$ and of $\triangle PCD$ (both from the vertex P) be h_1 and h_2, respectively. Because ΔPAB and ΔPCD are similar, we have

$$\begin{aligned} \frac{h_1}{h_2} &= \frac{\ell_1}{\ell_2} = \sqrt{\frac{36}{25}} = \frac{6}{5}. \quad \therefore h_2 = \frac{5}{6}h_1, \quad \ell_2 = \frac{5}{6}\ell_1. \\ \therefore [ABCD] &= \frac{1}{2}(\ell_1 + \ell_2) \cdot (h_1 + h_2) \\ &= \frac{1}{2}\left(\frac{11}{6}\ell_1\right) \cdot \left(\frac{11}{6}h_1\right) = \left(\frac{11}{6}\right)^2 \cdot \left(\frac{1}{2}\ell_1 h_1\right) \\ &= \left(\frac{11}{6}\right)^2 \cdot [PAB] = \left(\frac{11}{6}\right)^2 \cdot (36) = 11^2 = 121. \end{aligned}$$

Alternate Solution I. Let

$$\overline{AP} = x, \quad \overline{BP} = y, \quad \overline{CP} = u, \quad \overline{DP} = v, \quad \angle APB = \angle CPD = \theta.$$

Because $AB \parallel CD$, we have

$$\begin{aligned} [PAB] + [PBC] &= [ABC] = [ABD] \\ &= [PAB] + [PDA]. \\ \therefore\ [PBC] = [PDA] &= \frac{1}{2}yu\sin(\pi - \theta) = \frac{1}{2}xv\sin(\pi - \theta) \\ &= \sqrt{\left(\frac{1}{2}xy\sin\theta\right) \cdot \left(\frac{1}{2}uv\sin\theta\right)} \quad \text{(Note: } \sin\theta > 0) \\ &= \sqrt{[PAB] \cdot [PCD]} \\ &= \sqrt{36 \cdot 25} = 6 \cdot 5 = 30. \\ \therefore\ [ABCD] &= 36 + 2 \cdot 30 + 25 = (6+5)^2 = 121\ cm^2. \end{aligned}$$

Alternate Solution II. As shown in Alternate Solution I, we have $[PBC] = [PDA]$. Because ΔPAB and ΔPBC, also ΔPDA and ΔPCD have, respectively, the same height over the base line AC,

$$\frac{[PAB]}{[PBC]} = \frac{\overline{PA}}{\overline{PC}} = \frac{[PDA]}{[PCD]} = \frac{[PBC]}{[PCD]}.$$

$$\begin{aligned}\therefore\ [PBC]^2 &= [PDA]^2 = [PAB]\cdot[PCD] = 25\cdot 36.\\ [PBC] &= [PDA] = 5\cdot 6 = 30.\\ [ABCD] &= [PAB]+[PBC]+[PCD]+[PDA]\\ &= 36+30+25+30 = 121\ cm^2.\end{aligned}$$

148. The minimum possible area is 81 cm^2.

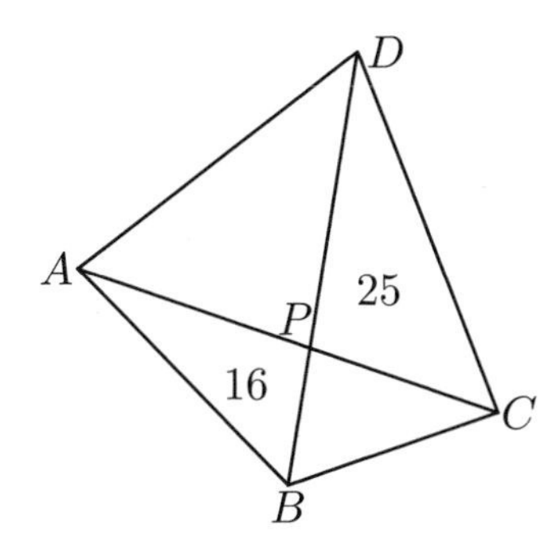

Figure 148

Let $\overline{AP}=x,\ \overline{BP}=y,\ \overline{CP}=u,\ \overline{DP}=v,\ \angle APB=\angle CPD=\theta$. Then

$$[PAB]=\frac{1}{2}xy\sin\theta=16,\quad [PCD]=\frac{1}{2}uv\sin\theta=25.$$

Using the inequality that the arithmetic mean is never less than the geometric mean:

$$\frac{1}{2}(a+b)\geq\sqrt{ab}\qquad (a\geq 0,\ b\geq 0),$$

(with equality holding if and only if $a=b$), we obtain

$$\begin{aligned}[PBC]+[PDA] &= \frac{1}{2}yu\sin(\pi-\theta)+\frac{1}{2}xv\sin(\pi-\theta)\\ &= \frac{1}{2}(yu+xv)\sin\theta\\ &\geq \sqrt{(yu)\cdot(xv)}\sin\theta\\ &= 2\sqrt{\left(\frac{1}{2}xy\sin\theta\right)\cdot\left(\frac{1}{2}uv\sin\theta\right)}\qquad (\text{Note: }\ \sin\theta>0)\\ &= 2\sqrt{16\cdot 25}=2\cdot 4\cdot 5.\\ \therefore\ [ABCD] &= [PAB]+\{[PBC]+[PDA]\}+[PCD]\\ &\geq 16+2\cdot 4\cdot 5+25=(4+5)^2=81\ cm^2.\end{aligned}$$

The minimum area 81 cm^2 is attained if and only if $\triangle PBC$ and $\triangle PDA$ have the same area (i.e., when $xv=yu$); in particular, when the quadrangle is a trapezoid.

Alternate Solution.

$$\begin{aligned}\frac{[PBC]}{[PAB]} &= \frac{\overline{PC}}{\overline{PA}}=\frac{[PCD]}{[PDA]}.\\ \therefore\ [PBC]\cdot[PDA] &= [PAB]\cdot[PCD]=16\cdot 25.\end{aligned}$$

By the inequality between the arithmetic mean and the geometric mean,

$$\begin{aligned}[PBC]+[PDA] &\geq 2\sqrt{[PBC]\cdot[PDA]}\\ &= 2\sqrt{[PAB]\cdot[PCD]}\\ &= 2\sqrt{16\cdot 25}=2\cdot 4\cdot 5=40.\\ \therefore\ [ABCD] &= ([PAB]+[PCD])+([PBC]+[PDA])\\ &\geq (16+25)+40=81\ cm^2.\end{aligned}$$

Remark. The trapezoid is not the only case attaining the minimum area. Can you find another such example?

Question: What can be said about the maximum possible area of quadrangle $ABCD$?

149. $\overline{AB}=2\sqrt{3}$.

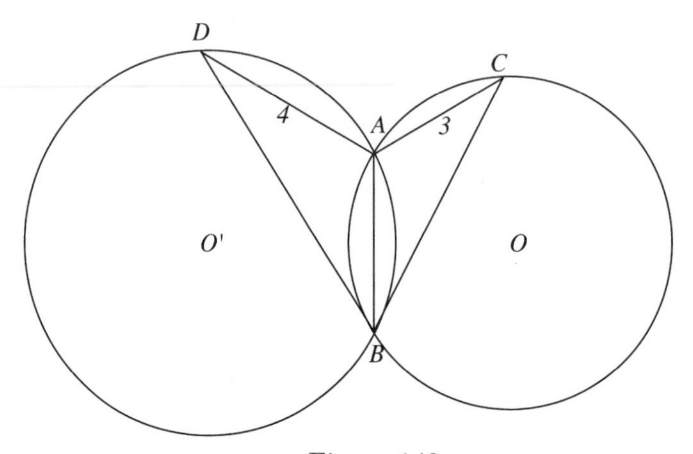

Figure 149

Because BC is tangent to the circle O', we have $\angle ABC=\angle ADB$. (See L.-s. Hahn: *Complex Numbers and Geometry,* Mathematical Association of America, Washington, D.C., 1994, p.180.) Similarly, because BD is tangent to the circle O, we have $\angle ACB=\angle ABD$. It follows that

$$\begin{aligned}\triangle ABC &\sim \triangle ADB.\quad \therefore\ \frac{\overline{AB}}{\overline{AC}}=\frac{\overline{AD}}{\overline{AB}}.\\ \overline{AB}^2 &= \overline{AC}\cdot\overline{AD}=3\cdot 4.\quad \therefore\ \overline{AB}=2\sqrt{3}.\end{aligned}$$

150. $\angle A=\dfrac{\pi}{3}$ or $\dfrac{2\pi}{3}$.

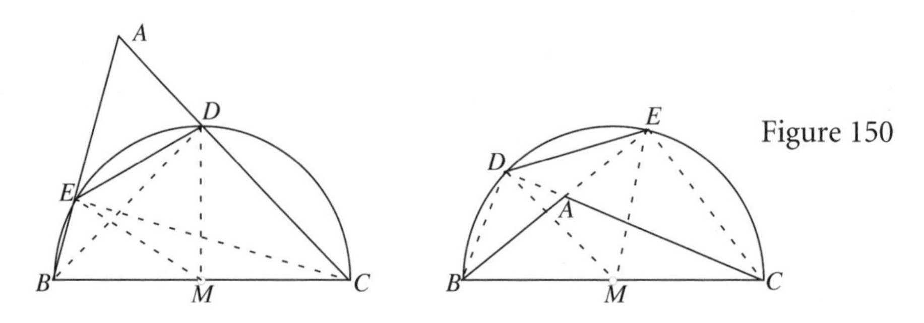

Figure 150

Because the points D, E are on the circle with BC as a diameter,

$$\overline{MD} = \overline{ME} = \frac{1}{2}\overline{BC}.$$

Hence it is sufficient to find the condition that $\overline{DE} = \frac{1}{2}\overline{BC}$. Now, by the law of sines

$$\frac{\overline{DE}}{\sin(\angle DBE)} = \overline{BC}.$$

But

$$\sin(\angle DBE) = \sin\left(\frac{\pi}{2} - \angle BAC\right) = \cos(\angle BAC)$$

if the vertex A is outside the circle whose diameter is BC; and

$$\sin(\angle DBE) = \sin\left(\angle BAC - \frac{\pi}{2}\right) = -\cos(\angle BAC)$$

if the vertex A is inside the circle whose diameter is BC. Hence the given condition is satisfied if and only if

$$\cos(\angle BAC) = \pm\frac{1}{2}; \text{ i.e., } \angle BAC = \frac{\pi}{3} \text{ or } \frac{2\pi}{3}.$$

(Note that the given condition cannot be satisfied when the vertex A is on the circle whose diameter is BC, because in this case $\angle BAC = \frac{\pi}{2}$ and both D and E coincide with A; i.e., $\overline{DE} = 0$.)

151. (a) A circular arc. (b) $2\sqrt{3}\pi$ cm.

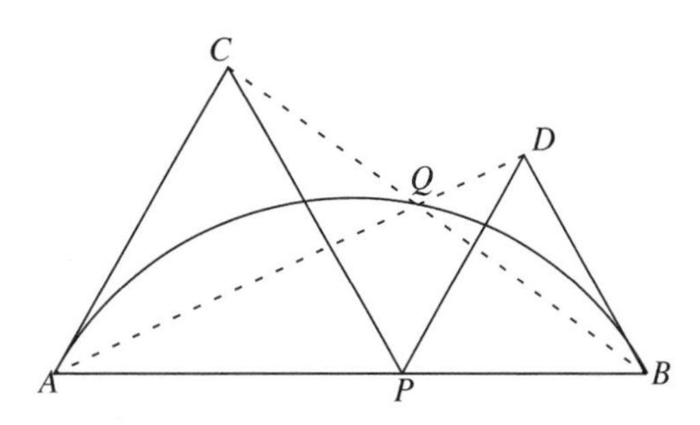

Figure 151

(a) Clearly, rotating $\triangle APD$ around P by $\frac{\pi}{3}$, we obtain $\triangle CPB$.

$$\therefore \angle AQB = \pi - \angle BQD = \pi - \frac{\pi}{3} = \frac{2\pi}{3}.$$

Therefore, Q is on the circular arc on which the chord AB subtends an angle $\frac{2\pi}{3}$. In other words, let $\triangle ABE$ be an equilateral triangle drawn on the opposite side of C,

D with respect to AB. Then the locus of Q is $\dfrac{1}{3}$ of the circumcircle of $\triangle ABE$ that lies on the same side of C, D with respect to AB.

(b) Because $r \cdot \cos\dfrac{\pi}{6} = \dfrac{1}{2}\overline{AB}$, where r is the radius of the circumcircle of $\triangle ABE$, $r = \dfrac{2}{\sqrt{3}} \cdot \dfrac{9}{2} = 3\sqrt{3}$, and the length of the locus is $\dfrac{1}{3} \cdot 2\pi(3\sqrt{3}) = 2\sqrt{3}\pi$.

Question: What if point P moves outside line segment AB but still on line AB? What if P is not on line AB?

152. $\dfrac{2ab}{a+b} < \sqrt{ab} < \dfrac{a+b}{2} < \sqrt{\dfrac{a^2+b^2}{2}}$.

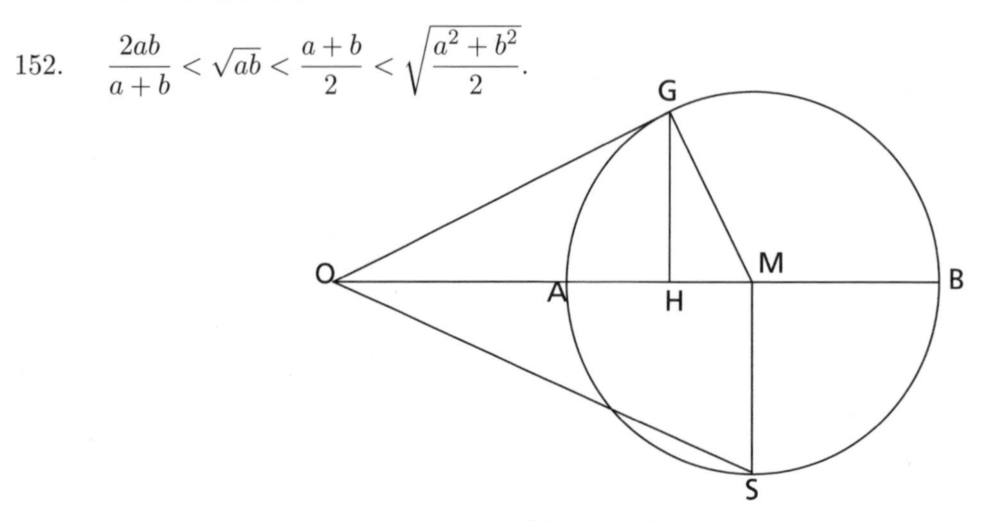

Figure 152

Because M is the midpoint of AB, we have

$$\begin{aligned}
\overline{OM} &= \frac{1}{2}(\overline{OA}+\overline{OB}) = \frac{a+b}{2}, \\
\overline{MS} &= \overline{MG} = \frac{1}{2}\overline{AB} = \frac{b-a}{2}, \\
\overline{OS}^2 &= \overline{OM}^2 + \overline{MS}^2. \\
\therefore\ \overline{OS} &= \sqrt{\left(\frac{a+b}{2}\right)^2 + \left(\frac{b-a}{2}\right)^2} = \sqrt{\frac{a^2+b^2}{2}}.
\end{aligned}$$

Similarly,

$$\overline{OG} = \sqrt{\overline{OM}^2 - \overline{MG}^2} = \sqrt{\left(\frac{a+b}{2}\right)^2 - \left(\frac{b-a}{2}\right)^2} = \sqrt{ab}.$$

Alternatively,

$$\overline{OG}^2 = \overline{OA} \cdot \overline{OB}. \quad \therefore\ \overline{OG} = \sqrt{ab}.$$

It remains to find $\overline{OH}$. Because $\triangle OHG \sim \triangle OGM$,

$$\frac{\overline{OH}}{\overline{OG}} = \frac{\overline{OG}}{\overline{OM}}. \quad \therefore\ \overline{OH} = \frac{\overline{OG}^2}{\overline{OM}} = \frac{2ab}{a+b}.$$

Remark. Given two positive real numbers a and b,

$$\frac{2ab}{a+b}, \quad \sqrt{ab}, \quad \frac{a+b}{2}, \quad \sqrt{\frac{a^2+b^2}{2}},$$

are called their *harmonic mean*, *geometric mean*, *arithmetic mean*, and *root mean square*, respectively. Note that the reciprocal of the harmonic mean is the arithmetic mean of the reciprocals (of a and b):

$$\left(\frac{2ab}{a+b}\right)^{-1} = \frac{1}{2}\left(\frac{1}{a}+\frac{1}{b}\right).$$

Exercise. (a) Show that for any three positive real numbers a, b, c which are not all equal, we have

$$\frac{3abc}{bc+ca+ab} < \sqrt[3]{abc} < \frac{a+b+c}{3} < \sqrt{\frac{a^2+b^2+c^2}{3}} < \sqrt[3]{\frac{a^3+b^3+c^3}{3}}.$$

(b) Can you generalize further?

153. The diameter is 9.

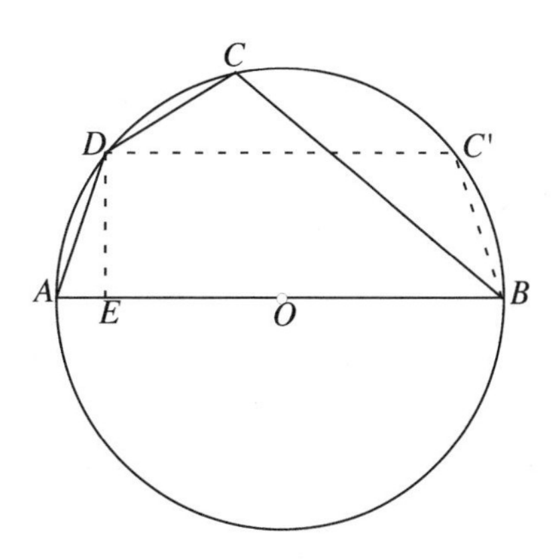

Figure 153(a)

Solution I. Note that we may move C to C' on the arc $\overset{\frown}{BC}$ so that

$$\overline{BC'} = 3 \quad \text{and} \quad \overline{C'D} = 7.$$

Then $ABC'D$ is a trapezoid. Let O be the center of the circle, E the foot of the perpendicular from D to the diameter AB, r the radius of the circle. Then $\overline{OE} = \frac{1}{2}\overline{C'D} = \frac{7}{2}$, and we have, by the Pythagorean theorem,

$$\begin{aligned} \overline{AD}^2 - \overline{AE}^2 &= \overline{DE}^2 = \overline{OD}^2 - \overline{OE}^2. \\ \therefore\ 3^2 - \left(r - \frac{7}{2}\right)^2 &= r^2 - \left(\frac{7}{2}\right)^2. \\ 2r^2 - 7r - 9 &= (r+1)(2r-9) = 0. \end{aligned}$$

Hence the diameter is $2r = 9$, because $r > 0$.

Solution II. Let $\overline{AC'} = \overline{BD} = x$, $\overline{AB} = d$. Then because $\angle ADB = \frac{\pi}{2}$, we have

$$d^2 = x^2 + 3^2, \quad x^2 = 7d + 3^2,$$

where the second equality follows from the Ptolemy theorem (applied to the quadrangle $ABC'D$): For a cyclic quadrangle, the product of the lengths of two diagonals is equal to the sum of the products of the lengths of the pairs of the opposite sides. (See L.-s. Hahn: *Complex Numbers and Geometry,* Mathematical Association of America, Washington, D.C., 1994, p.64.) Eliminating x^2 between these two equalities, we obtain

$$d^2 - 7d - 18 = (d + 2)(d - 9) = 0.$$

Only $d = 9$ is a valid answer.

Solution III. Let $\overline{BD} = x$, $\overline{AC} = y$. Then

$$d^2 = x^2 + 3^2 = y^2 + 7^2, \quad xy = 3d + 3 \cdot 7,$$

where the last equality follows from the Ptolemy theorem (applied to the quadrangle $ABCD$). Solving for x and y in terms of d from the first equation and substituting into (the square of) the second equation, we obtain

$$(d^2 - 9)(d^2 - 49) = 3^2 \cdot (d + 7)^2.$$

Rewriting this equation, we obtain

$$d(d^3 - 67d - 126) = d(d + 2)(d + 7)(d - 9) = 0.$$

Only $d = 9$ is a valid answer.

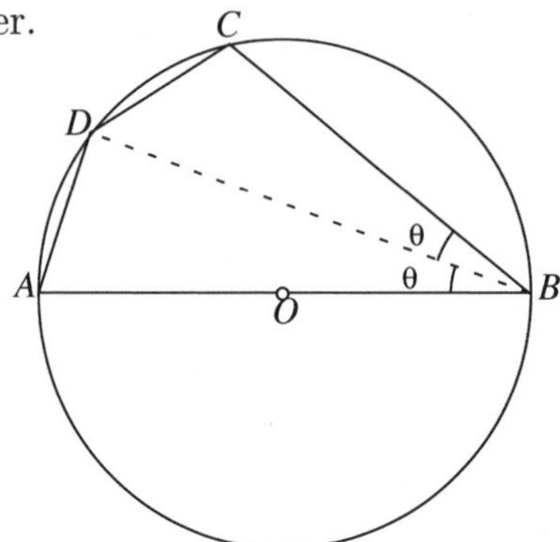

Figure 153(b)

Trigonometric Solution. Let $\angle CBD = \angle DBA = \theta$. Then

$$\sin\theta = \frac{\overline{DA}}{\overline{AB}} = \frac{3}{d},$$

$$\cos 2\theta = \frac{\overline{BC}}{\overline{AB}} = \frac{7}{d}.$$

But $\cos 2\theta = 1 - 2\sin^2\theta$.

$$\therefore \frac{7}{d} = 1 - 2\left(\frac{3}{d}\right)^2, \quad d^2 - 7d - 18 = (d + 2)(d - 9) = 0. \quad \therefore d = 9.$$

154. 60 Pascal lines.

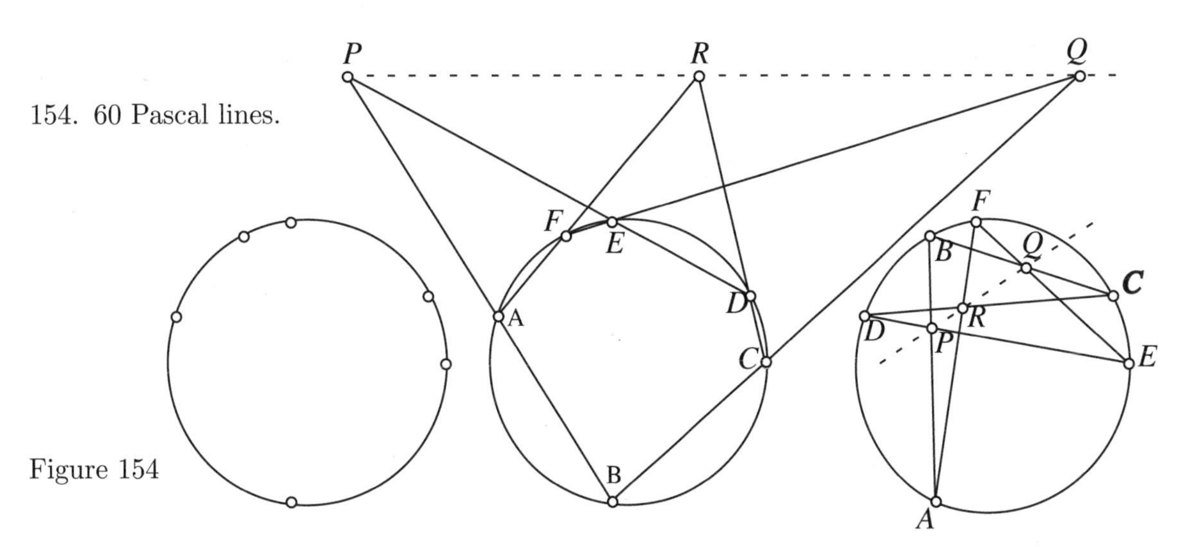

Figure 154

Given a 'hexagon' (which may be self-intersecting), we can label any one of its six vertices as A, and either of the two adjacent vertices as B. Now, the labels for the rest of the vertices are determined uniquely in alphabetical order. Thus, there are $6 \cdot 2 = 12$ ways to label the hexagon. Because there are 6! ways to label six points, we have $\frac{6!}{12} = 60$ Pascal lines.

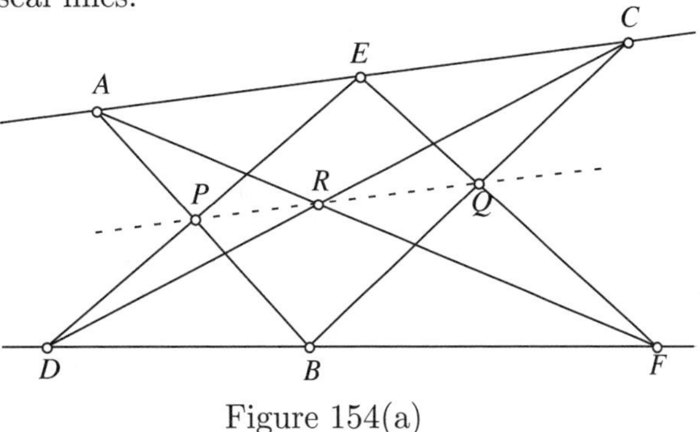

Figure 154(a)

Exercise. Suppose A, C, E are three points on one line, and B, D, F on another. Let P, Q, R denote the intersection points (of the extensions if necessary) of the 'opposite sides' AB and DE, BC and EF, CD and FA, respectively (assuming they all intersect). A theorem of Pappus says that points P, Q, and R lie on a line. Let us call this line a ***Pappus line*** of the 'hexagon' $ABCDEF$. What is the maximum possible number of Pappus lines (for the same six points) by changing the order of labels?

155. $\overline{AP} = 2$.

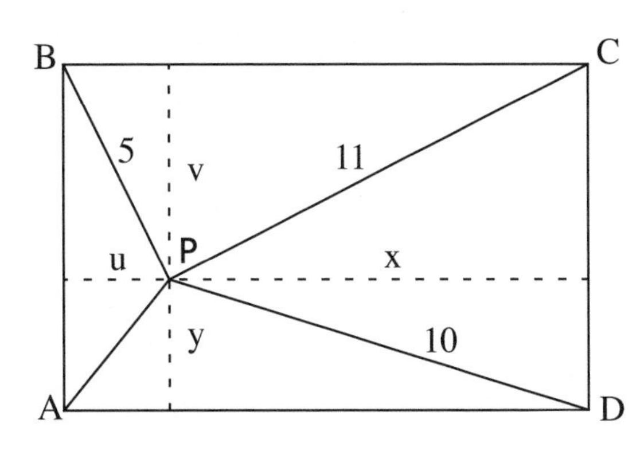

Figure 155

Let u, v, x, y be the lengths of the perpendiculars from P to AB, BC, CD, DA, respectively. Then

$$\overline{AP}^2 = y^2 + u^2, \qquad \overline{BP}^2 = u^2 + v^2,$$
$$\overline{CP}^2 = v^2 + x^2, \qquad \overline{DP}^2 = x^2 + y^2.$$

$$\begin{aligned} \therefore\ \overline{AP}^2 + \overline{CP}^2 &= (y^2 + u^2) + (v^2 + x^2) \\ &= (u^2 + v^2) + (x^2 + y^2) \\ &= \overline{BP}^2 + \overline{DP}^2. \\ \therefore\ \overline{AP}^2 &= \overline{BP}^2 + \overline{DP}^2 - \overline{CP}^2 \\ &= 5^2 + 10^2 - 11^2 = 4 = 2^2. \\ \overline{AP} &= 2. \end{aligned}$$

Remark. Note carefully that our computation is valid even if the point P is outside the rectangle $ABCD$. In fact, the result is correct even if P is not in the plane of the rectangle $ABCD$.

156. (a) $\angle BPC = \frac{\pi}{2}$, $\angle CPA = \frac{5\pi}{6}$, $\angle APB = \frac{2\pi}{3}$. (b) $\overline{BC} = \sqrt{7}\ cm$.

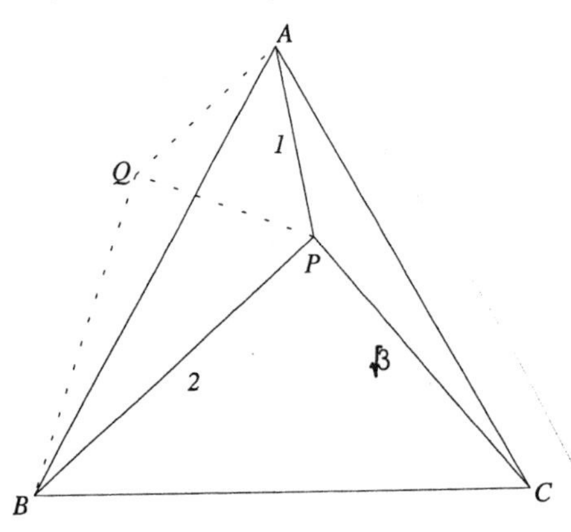

Figure 156(a)

(a) Construct an equilateral triangle APQ as in Figure 156(a). Now rotate $\triangle ACP$ around A by $\frac{\pi}{3}$; we see that $\triangle ACP \cong \triangle ABQ$, and so $\overline{BQ} = \overline{CP} = \sqrt{3}$. Therefore, the three sides BP, PQ, QB of $\triangle BPQ$ have length 2, 1, $\sqrt{3}$, respectively. Hence $\triangle BPQ$ is a 30^0-60^0-90^0 triangle. It follows that

$$\begin{aligned} \angle APB &= \angle APQ + \angle QPB = \frac{\pi}{3} + \frac{\pi}{3} = \frac{2\pi}{3}, \\ \angle CPA &= \angle BQA = \angle BQP + \angle PQA = \frac{\pi}{2} + \frac{\pi}{3} = \frac{5\pi}{6}, \\ \therefore\ \angle BPC &= 2\pi - (\angle APB + \angle CPA) \\ &= 2\pi - \left(\frac{2\pi}{3} + \frac{5\pi}{6}\right) = \frac{\pi}{2}. \end{aligned}$$

(b) By the Pythagorean theorem, we have

$$\overline{BC}^2 = \overline{PB}^2 + \overline{PC}^2 = 2^2 + (\sqrt{3})^2 = 7. \quad \therefore \overline{BC} = \sqrt{7}\ cm.$$

Question: What if point P is outside equilateral triangle ABC?

157. (a) $\angle APB = \frac{3\pi}{4}$. (b) $\overline{AB} = \sqrt{29}$.

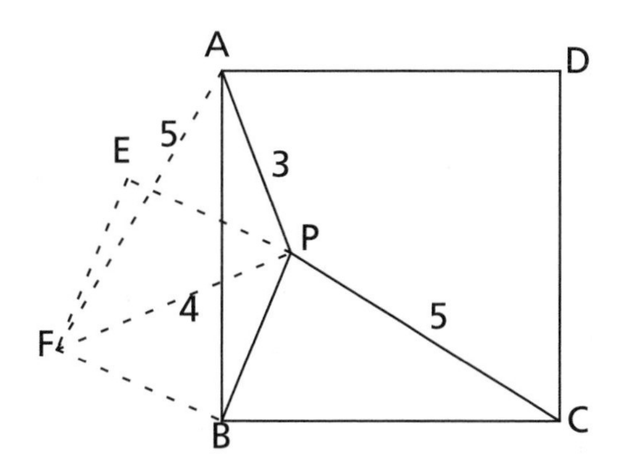

Figure 157(a)

(a) Construct the square $BPEF$ with PB as one of its sides as in Figure 157(a). Then it is simple to see that when $\triangle BCP$ is rotated by $\frac{\pi}{2}$ around the point B, it coincides with $\triangle BAF$. Therefore, $\overline{AF} = \overline{CP} = 5$. Thus the three sides of $\triangle AFP$ are 3, 4, 5, respectively.

$$\therefore \angle APB = \angle APF + \angle FPB = \frac{\pi}{2} + \frac{\pi}{4} = \frac{3\pi}{4}.$$

(b) By the law of cosines, we have

$$\begin{aligned} \overline{AB}^2 &= \overline{AP}^2 + \overline{BP}^2 - 2\overline{AP} \cdot \overline{BP} \cdot \cos(\angle APB) \\ &= 3^2 + (2\sqrt{2})^2 - 2 \cdot 3 \cdot 2\sqrt{2} \cdot \cos\frac{3\pi}{4}. \\ &= 9 + 8 + 12\sqrt{2} \cdot \frac{\sqrt{2}}{2} = 29. \\ \therefore \overline{AB} &= \sqrt{29}. \end{aligned}$$

Question: What if point P is outside square $ABCD$?

158. (a) $a = 2$, $b = 1$. (b) $c = 2$, $d = 3$. (c) $p = 3$, $q = 2$, $r = 2$, $s = 1$.

(a) In Figure 158, $\overline{AB} = \overline{BC} = 1$, $\overline{AC} = \overline{CD} = \sqrt{2}$, and so

$$\begin{aligned} \angle ADB &= \frac{1}{2}\angle ACB = \frac{\pi}{8}. \\ \therefore \tan\frac{\pi}{8} &= \frac{\overline{AB}}{\overline{BD}} = \frac{1}{\sqrt{2}+1} = \sqrt{2} - 1. \end{aligned}$$

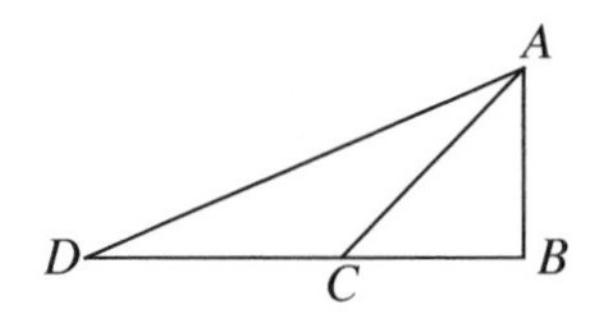

Figure 158

(b) In Figure 158(a), let $\overline{EF} = 1$, $\overline{FG} = \sqrt{3}$, $\overline{EG} = \overline{GH} = 2$, $\angle EFG = \dfrac{\pi}{2}$. Then

$$\begin{aligned}\angle EHF &= \frac{1}{2}\angle EGF = \frac{\pi}{12}.\\ \therefore\ \tan\frac{\pi}{12} &= \frac{\overline{EF}}{\overline{FH}} = \frac{1}{\sqrt{3}+2} = 2-\sqrt{3}.\end{aligned}$$

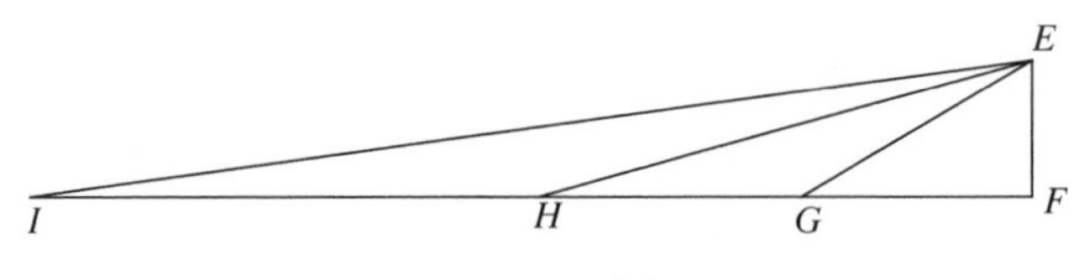

Figure 158(a)

(c) In Figure 158(a),

$$\overline{HI} = \overline{EH} = \sqrt{1^2 + (2+\sqrt{3})^2} = \sqrt{8+4\sqrt{3}} = \sqrt{6}+\sqrt{2}.$$

(The last step is by the technique used in the solution of Problem 37.) Then

$$\begin{aligned}\angle EIF &= \frac{1}{2}\angle EHF = \frac{\pi}{24}.\\ \therefore\ \tan\frac{\pi}{24} &= \frac{\overline{EF}}{\overline{FI}} = \frac{1}{\sqrt{3}+2+\sqrt{6}+\sqrt{2}}\\ &= \frac{1}{(\sqrt{3}+\sqrt{2})(\sqrt{2}+1)} = (\sqrt{3}-\sqrt{2})(\sqrt{2}-1).\end{aligned}$$

159. (a) 12-gon (dodecagon). (b) 24 (square units).

Observe that

$$\Big||x|-1\Big| + \Big||y|-1\Big| \le 2$$

remains unchanged if we replace x by $-x$ and/or y by $-y$. It follows that the polygon must be symmetric with respect to both coordinate axes (and the origin). Therefore, if we know the portion of the polygon in the first quadrant, then we know the total picture of the polygon by reflections. Now in the first quadrant, $|x| = x$ and $|y| = y$, and hence the given inequality becomes

$$|x-1| + |y-1| \le 2.$$

But this is just the parallel translation of the square

$$|x| + |y| \le 2$$

(with center at the origin, and area 8), to the congruent square with center at (1, 1). Hence the polygon is as in the figure on the right.

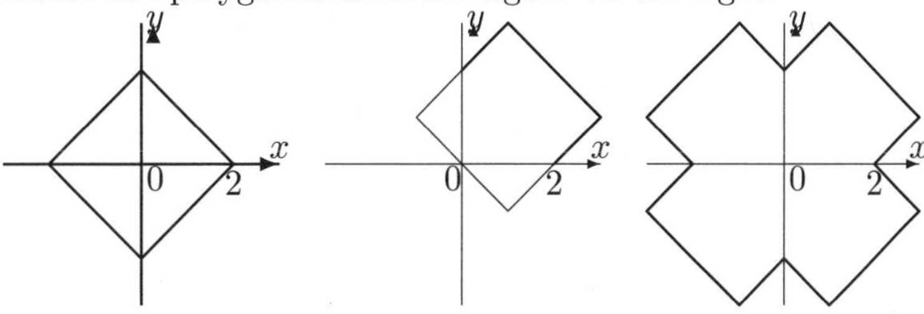

Figure 159(a)

Knowing the picture, the answers are immediate.

160. These kinds of problems are best handled using complex numbers. [See author's book: *Complex Numbers and Geometry* (Mathematical Association of America, 1994.)]

Let O be the origin of the complex plane and $A(\alpha)$, $C(\gamma)$, $E(\epsilon)$. Then $B(\tau\alpha)$, $D(\tau\gamma)$, $F(\tau\epsilon)$, where $\tau = \cos\frac{\pi}{3} + i\sin\frac{\pi}{3}$ $(= (1+i\sqrt{3})/2 = e^{i\pi/3})$, and

$$U\left(\frac{\tau\gamma+\epsilon}{2}\right), \quad V\left(\frac{\tau\epsilon+\alpha}{2}\right), \quad W\left(\frac{\tau\alpha+\gamma}{2}\right).$$

Thus

$$\begin{aligned} \overrightarrow{UV} &= \frac{1}{2}\{\alpha - \tau\gamma + (\tau-1)\epsilon\}, \\ \overrightarrow{UW} &= \frac{1}{2}\{\tau\alpha + (1-\tau)\gamma - \epsilon\}. \end{aligned}$$

To show $\triangle UVW$ is equilateral, it is sufficient to show that rotating $\overrightarrow{UV}$ by $\pi/3$ (i.e., multiplying by τ), we obtain $\overrightarrow{UW}$:

$$\begin{aligned} \tau\overrightarrow{UV} &= \frac{\tau}{2}\{\alpha - \tau\gamma + (\tau-1)\epsilon\} \\ &= \frac{1}{2}\{\tau\alpha - \tau^2\gamma + (\tau^2-\tau)\epsilon\} \\ &= \frac{1}{2}\{\tau\alpha + (1-\tau)\gamma - \epsilon\} \\ &= \overrightarrow{UW}, \end{aligned}$$

where we used $\tau^2 - \tau + 1 = 0$.
The case of $\triangle XYZ$ is proved in the same manner.

Exercise. Prove the following generalization:
Suppose $\triangle OAB \sim \triangle DOC \sim EFO$; i.e., these triangles are similar with the same orientation.

(a) Show that $\triangle UVW$ is a fourth similar triangle, where U, V, W are the midpoints of DE, FA, BC, respectively.

(b) What can be said about $\triangle XYZ$, where X, Y, Z are the midpoints of CF, EB, AD, respectively?

Hint: Apply Exercise 4 in Chapter 2 of *Complex Numbers and Geometry.*

161. (a) $k = -3$. (b) $x^2 - 3y^2 = 1$.

(a) Substituting $x_{n+1} = 2x_n + 3y_n$, $y_{n+1} = x_n + 2y_n$ into the left-hand side of

$$x_{n+1}^2 + ky_{n+1}^2 = x_n^2 + ky_n^2,$$

we obtain

$$\begin{aligned}&(2x_n + 3y_n)^2 + k\,(x_n + 2y_n)^2 \\ &\quad = (4+k)x_n^2 + (12+4k)x_ny_n + (9+4k)y_n^2.\end{aligned}$$

Comparing the coefficients, we obtain

$$4 + k = 1, \quad 12 + 4k = 0, \quad 9 + 4k = k.$$

This over-determined system of simultaneous equations has a solution $k = -3$.

$$\therefore\ x_{n+1}^2 - 3y_{n+1}^2 = x_n^2 - 3y_n^2.$$

(b) From our computations in Part (a),

$$x_{n+1}^2 - 3y_{n+1}^2 = x_n^2 - 3y_n^2 = x_{n-1}^2 - 3y_{n-1}^2 = \cdots = x_0^2 - 3y_0^2 = 1.$$

Hence all the points (x_n, y_n) are on the hyperbola $x^2 - 3y^2 = 1$.

Remark. Note that $x_n + y_n\sqrt{3} = \left(2+\sqrt{3}\right)^n$.

162. $x^2 - 2y^2 = 1$.

$$\begin{aligned}x_n + y_n\sqrt{2} &= \left(3+2\sqrt{2}\right)^n, \quad x_n - y_n\sqrt{2} = \left(3-2\sqrt{2}\right)^n. \\ \therefore\ x_n^2 - 2y_n^2 &= \left(x_n + y_n\sqrt{2}\right)\cdot\left(x_n - y_n\sqrt{2}\right) \\ &= \left\{\left(3+2\sqrt{2}\right)\cdot\left(3-2\sqrt{2}\right)\right\}^n = 1.\end{aligned}$$

163. $2x - y = 0$ and $x + y = 0$.

Suppose $ax + by + c = 0$ is the equation of a line satisfying the condition. Then

$$a(x+2y) + b(4x+3y) + c = 0;\ \text{i.e.,}\ (a+4b)x + (2a+3b)y + c = 0$$

must be the same line as $ax + by + c = 0$. If $c \neq 0$, then we must have

$$a + 4b = a, \ \text{ and }\ 2a + 3b = b. \quad \therefore\ a = b = 0.$$

Hence this case cannot happen. So $c = 0$. But then we obtain

$$\begin{aligned} \frac{a+4b}{a} &= \frac{2a+3b}{b}; \text{ i.e., } (a+4b)b = a(2a+3b). \\ 2a^2+2ab-4b^2 &= 2(a+2b)(a-b) = 0. \quad \therefore a = -2b \text{ or } a = b. \end{aligned}$$

Hence the equations of the lines with the given property must be

$$2x - y = 0 \quad \text{and} \quad x + y = 0.$$

It is simple to verify that these two lines indeed satisfy the required condition.

Alternate Solution (via linear algebra). Clearly, the matrix of the given transformation, which we shall call T, is

$$\begin{bmatrix} 1 & 2 \\ 4 & 3 \end{bmatrix}.$$

We want to find the invariant subspaces under T; in other words, we want to find the eigenspaces.

$$\det(T - \lambda I) = \begin{vmatrix} 1-\lambda & 2 \\ 4 & 3-\lambda \end{vmatrix} = 0$$

gives $(1-\lambda)(3-\lambda) - 8 = 0$, which simplifies to $\lambda^2 - 4\lambda - 5 = (\lambda+1)(\lambda-5) = 0$. If $\lambda = -1$, then the matrix becomes $\begin{bmatrix} 2 & 2 \\ 4 & 4 \end{bmatrix}$, so the associated eigenvectors are scalar muliples of $\begin{bmatrix} 1 \\ -1 \end{bmatrix}$; i.e., $x = t,\ y = -t$. Eliminating t, we obtain $x + y = 0$.

If $\lambda = 5$, then the matrix becomes $\begin{bmatrix} -4 & 2 \\ 4 & -2 \end{bmatrix}$, so the associated eigenvectors are scalar multiples of $\begin{bmatrix} 1 \\ 2 \end{bmatrix}$; i.e., $x = t,\ y = 2t$. Eliminating t, we obtain $2x - y = 0$.

164. We can construct the desired perpendicular line using a compass once and only once.

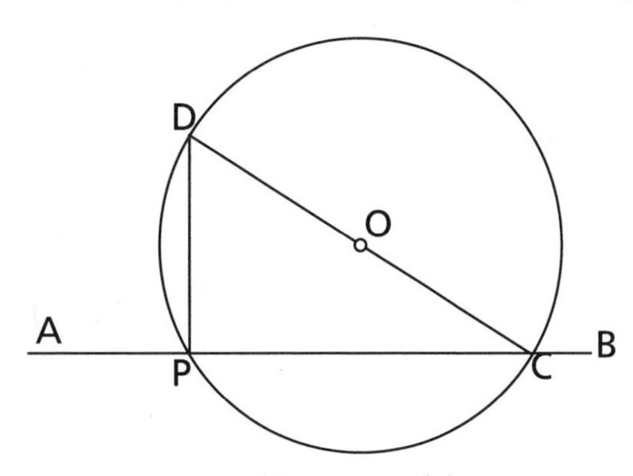

Figure 164(a)

Let P be the given point on line AB. Choose an arbitrary point O not on AB. Draw the circle with O as the center passing through P. Suppose the circle intersects line AB at C (other than P). Let CD be a diameter of the circle. Then DP is perpendicular to AB because $\angle CPD$ is subtended by the diameter CD.

If the circle does not intersect line AB at any point other than P, then O is already on the perpendicular line passing through P.

Remark. It is impossible to construct the desired perpendicular line without using a compass because an angle is not a «projective invariant.»

165. The problem suggests that there must be a relation between the lengths of projections of a segment and the side lengths of the triangle. Indeed the following theorem holds, and from which the answer $\overline{AB} = 7$, or 2 is an immediate consequence.

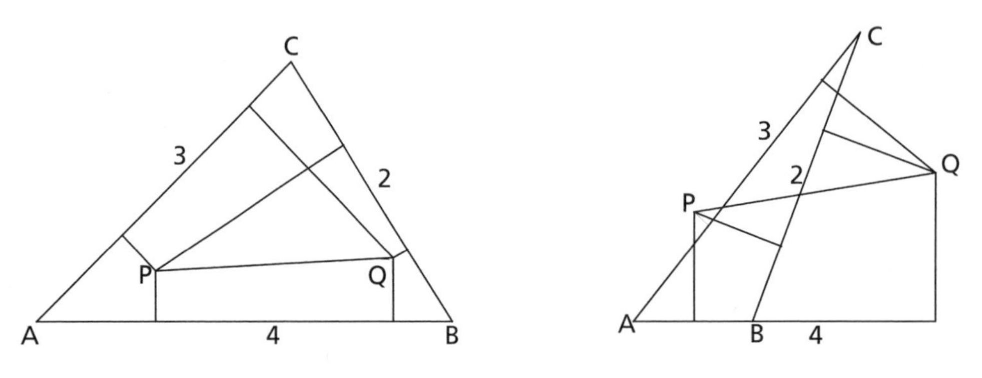

Figure 165(a)

Theorem. Suppose PQ is a line segment on the plane of $\triangle ABC$. Let the perpendicular projections of point P on (the extensions of) sides BC, CA, AB be D, E, F, and that of point Q be L, M, N. Then

$$\overline{BC} \cdot \overline{DL} + \overline{CA} \cdot \overline{EM} + \overline{AB} \cdot \overline{FN} = 0,$$

where $\overline{BC} \cdot \overline{DL} > 0$ if $\overrightarrow{BC}$ and $\overrightarrow{DL}$ have the same direction, but $\overline{BC} \cdot \overline{DL} < 0$ if they have the opposite direction, and $\overline{BC} \cdot \overline{DL} = 0$ if the points D and L coincide. Similarly, for the other products $\overline{CA} \cdot \overline{EM}$ and $\overline{AB} \cdot \overline{FN}$.

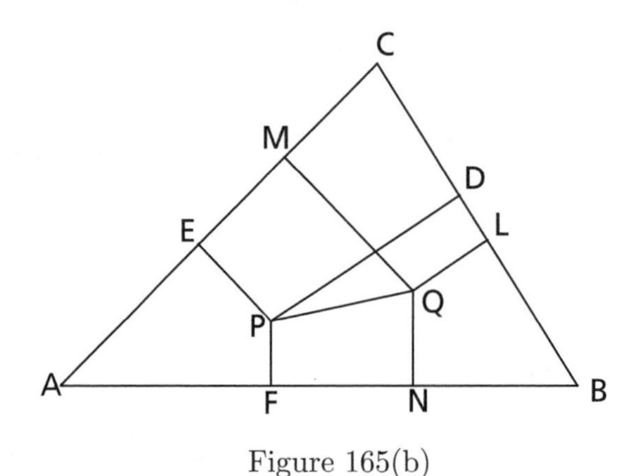

Figure 165(b)

Proof. Draw line segments UX, VY, and WZ that are perpendicular to PQ having the lengths twice that of PQ with A, B, C as their respective midpoints. Discard the portion that is inside $\triangle ABC$.

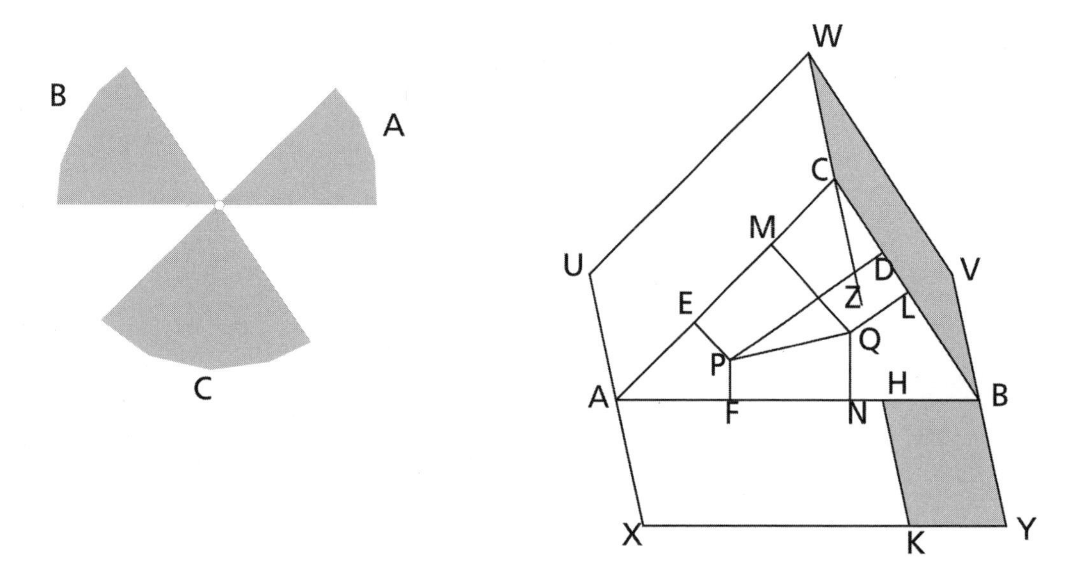

Figure 165(c)

Note that if PQ is not perpendicular to any of the three sides of $\triangle$ ABC, then there is bound to be one segment half of which is inside the triangle. This is easy to verify because the sum of the three angles of a triangle is π, and so if we parallel translate three angles so that their vertices meet at one point, then the union of the three angles covers half of the plane. Therefore, regardless of the direction of the line segment whose midpoint is at the common vertex of the angles, half of the segment must be in the half of the plane covered by the union of these three angles. (And the area of the parallelogram on the opposite side will be the sum of that of the other two.) The case that PQ is perpendicular to one of the three sides of a triangle is easy and is left for the reader.

Now it is easy to see that the height of parallelogram $BVWC$ on side BC is equal to the length of the projection DL of PQ on BC. Hence the absolute value of the product $\overline{BC} \cdot \overline{DL}$ is equal to the area of the parallelogram $BVWC$, which, in turn, is equal to that of parallelogram $BHKY$, where H, K are the intersections of the extension of WC with AB and XY, respectively. Similarly, the absolute value of the products $\overline{CA} \cdot \overline{EM}$ is equal to the area of parallelogram $CWUA$, which in turn is equal to that of parallelogram $AXKH$. The remainder of the proof is obvious and is left for the reader.

166. Let $A_1A_2 \cdots A_n$ be the convex polygon. Extend each of its sides, say counterclockwise, until they intersect the closed curve at B_1, B_2, $\cdots$, B_n, respectively, as in

Figure 166(a).

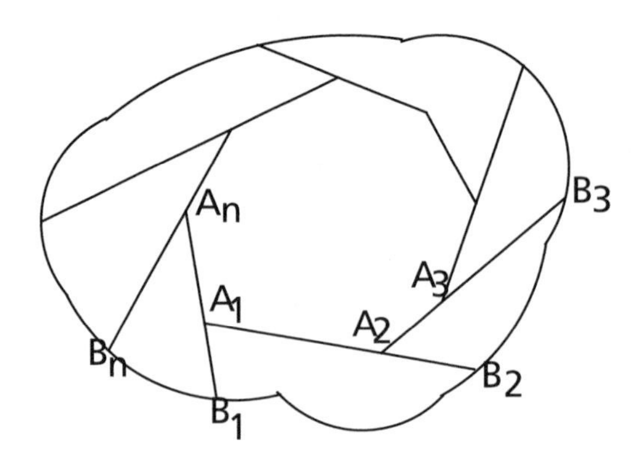

Figure 166(a)

Because the line segment gives the shortest path joining the endpoints, we have

$$\begin{aligned}
\overline{A_1A_2} + \overline{A_2B_2} = \overline{A_1B_2} &\le \overline{A_1B_1} + (B_1B_2),\\
\overline{A_2A_3} + \overline{A_3B_3} = \overline{A_2B_3} &\le \overline{A_2B_2} + (B_2B_3),\\
\cdots \quad \cdots &\le \cdots\\
\overline{A_nA_1} + \overline{A_1B_1} = \overline{A_nB_1} &\le \overline{A_nB_n} + (B_nB_1),
\end{aligned}$$

where (B_kB_{k+1}) denotes the length of the portion of the closed curve between B_k and B_{k+1} $(k = 1, 2, \cdots, n)$. Adding these inequalities, and canceling the terms $\overline{A_kB_k}$ $(k = 1, 2, \cdots, n)$ from both sides, we obtain

$$\overline{A_1A_2} + \overline{A_2A_3} + \cdots + \overline{A_nA_1} \le (B_1B_2) + (B_2B_3) + \cdots + (B_nB_1),$$

which is what we want to show.

Question: Where did we use the assumption that the inner polygon is convex? Also what if the convex polygon is replaced by a convex closed curve?

APPENDIX A

New Mexico Mathematics Contests

(Answers in Appendix B)

1990–1991 First Round (November 10, 1990)

1. It takes 70 minutes to travel by bus between Albuquerque and Santa Fe. If a bus company wants to run a bus every 20 minutes simultaneously from both places, what is the minimum number of buses needed?

2. Suppose a quadrilateral $ABCD$ circumscribes a circle, i.e., all four sides are tangent to the circle, and

$$\overline{BC} = 11\ cm, \quad \overline{CD} = 9\ cm, \quad \overline{DA} = 15\ cm.$$

What is the length $\overline{AB}$?

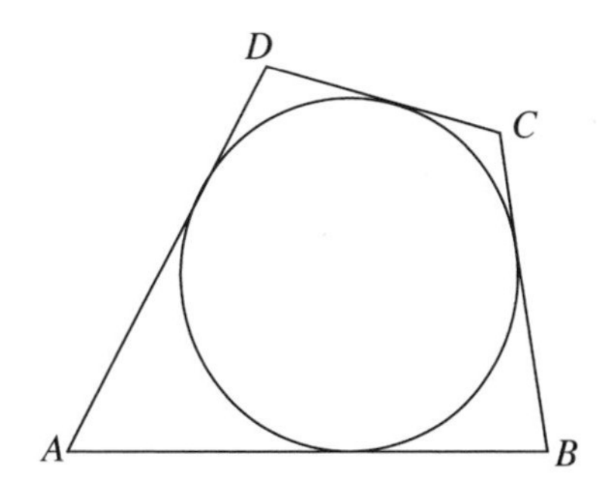

Figure A1

3. To prepare for the New Mexico Mathematics Contest, Albert, Beth and Chris decided to solve all the problems in their Geometry Problem Book. They all started on October 1. Albert managed to solve 8 problems per day (except possibly on the last day when there may have been fewer than 8 problems left), and finished solving all the problems in the book on November 2, while Beth solved 10 problems per day, (except, possibly, on the last day when there may have been fewer than 10 problems left), and finished on October 27. If Chris solved 7 problems per day (except, possibly, on the last day when there may have been fewer than 7 problems left), on what day will she finish solving all the problems in the book? (Note that October has 31 days!)

4. Find the sum of the 7 angles indicated in Figure A2.

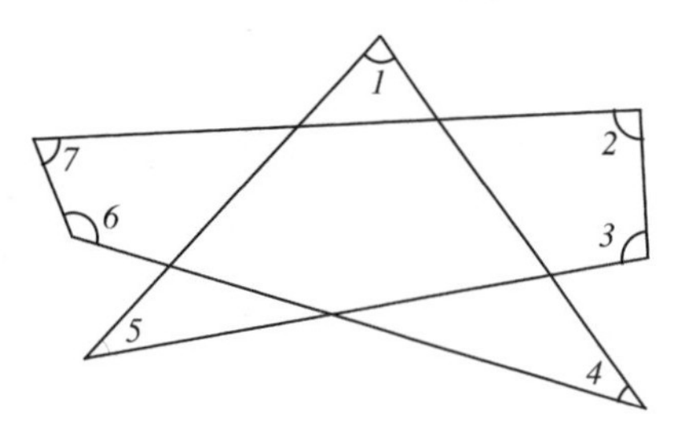

Figure A2

5. Upon graduating from U.N.M., David and Esther each get a job with a five-year contract. David's starting annual salary is \$30,000 with an annual raise of \$1,500 per year. Esther's starting monthly salary is \$2,500 with a monthly raise of \$10 per month. Who will earn more and by how much over the five-year period after graduation?

6. Let $\triangle ABC$ be a right triangle. Choose two points D and E on the hypotenuse AB such that

$$\overline{AD} = \overline{AC}, \quad \overline{BE} = \overline{BC}.$$

Suppose $\overline{AB} = 29\ cm$ and $\overline{DE} = 12\ cm$.
What are the lengths of the legs BC and AC?

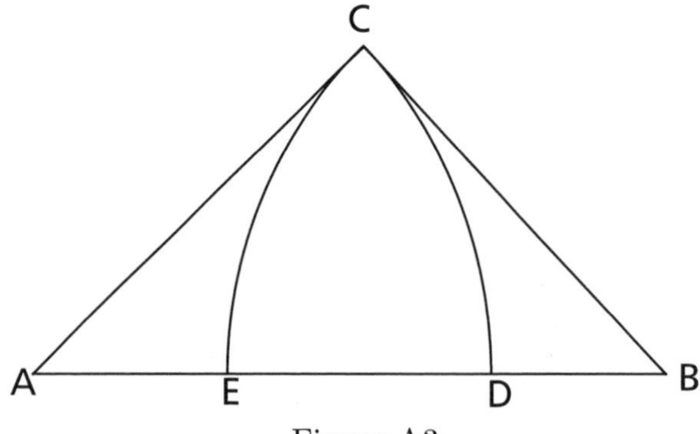

Figure A3

7. Suppose a, b, c, d are positive integers satisfying

$$a + \cfrac{1}{b + \cfrac{1}{c + \cfrac{1}{d}}} = \frac{1990}{991}.$$

Find the value of d.

8. Of the following 10 numbers, only one is a perfect square. Which one?

(a)	8344572651	(b)	7955896032
(c)	1695032253	(d)	4906358264
(e)	1782570645	(f)	5729581636
(g)	3213046377	(h)	2032918848
(i)	2973562479	(j)	3567659100

(This may look tedious and dull; try a shortcut.)

1990–1991 Final Round (February 2, 1991)

1. The distance between two points A and B is 18 *meters*, and there are poles every 60 *cm* between them. If we want to change the distance between each pair of neighboring poles from 60 *cm* to 75 *cm*, then what is the minimum number of poles we have to remove? (Notes: 1 *meter* = 100 *cm*, and originally there are 31 poles.)

2. Suppose $ABCD$ is a quadrilateral circumscribing a circle with center at O. If the areas of the triangles OAB and OCD are 12 cm^2 and 17 cm^2, respectively, what is the area of the quadrangle $ABCD$?

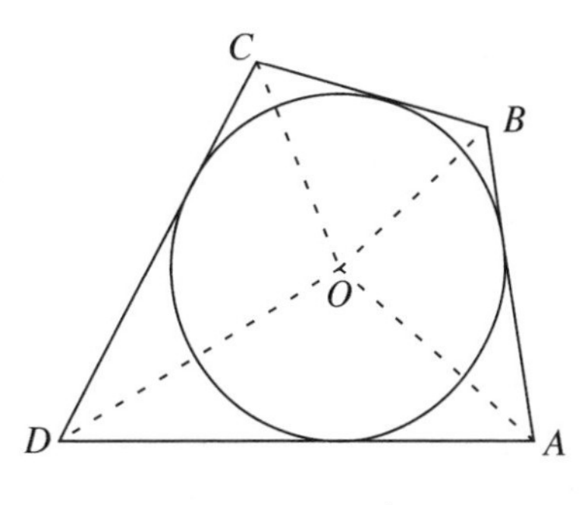

Figure A4

3. In dividing candies among children at a birthday party, 50 candies will be left over if we give 8 candies to each child. If we give 12 candies to each child, the last child will get less than one quarter of what the other children get. How many children are there and how many candies are there to give away?

4. A polygon that does not cross itself, and whose vertices all lie on lattice points is called a simple lattice polygon. (Lattice points are points whose coordinates are integers.) A theorem of Pick asserts that there are coefficients a, b, and c with the property that if H is a simple lattice polygon with p lattice points in its interior and q lattice points on its boundary (for example, in Figure A5, $p = 14$, and $q = 33$), then

$$\text{Area of } H = ap + bq + c.$$

In other words, it is possible to find the area of an arbitrary simple lattice polygon by just counting the lattice points. Assuming there are such coefficients, determine what these coefficients a, b, c must be.

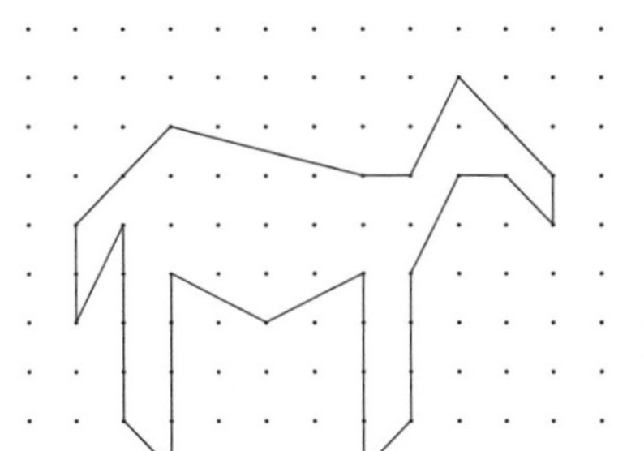

Figure A5

5. Can an integer with 2 or more digits, all of which are either 1, 3, 5, 7, or 9 (for example, 1991, 17, 731591375179, 753) be a perfect square? Justify your assertion.

6. On the sides of $\triangle ABC$, choose points P, Q, R, so that

$$\overline{BP} = \overline{PC}, \quad \overline{CQ} = 2\overline{QA}, \quad \overline{AR} = 2\overline{RB}.$$

Joining the points P, Q, R divides $\triangle ABC$ into four small triangles. If the areas of these four triangles are four consecutive integers (with cm^2 as the unit), what is the area of $\triangle ABC$?

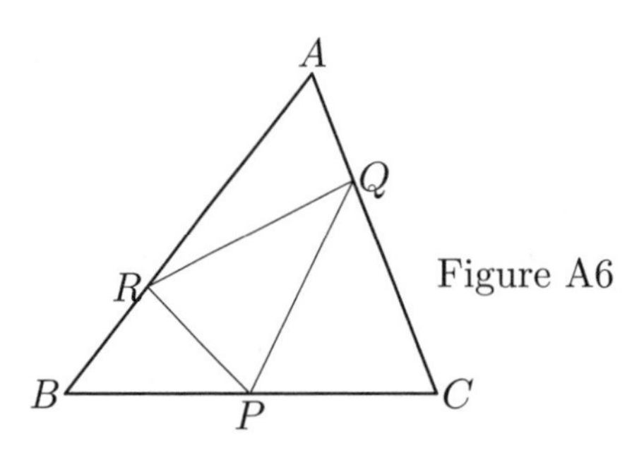

Figure A6

7. Find the integer n such that

$$n \leq \frac{1}{\sqrt{1}} + \frac{1}{\sqrt{2}} + \frac{1}{\sqrt{3}} + \cdots + \frac{1}{\sqrt{119}} + \frac{1}{\sqrt{120}} < n + 1.$$

(**Hint:** $\dfrac{1}{\sqrt{k} + \sqrt{?}} < \dfrac{1}{2\sqrt{k}} < \dfrac{1}{\sqrt{?} + \sqrt{k}}$; or if you wish you may use calculus.)

8. Does there exist an equilateral triangle (in the (x, y)-plane) all three of whose vertices are at lattice points? If your answer is Yes, give an example of such an equilateral triangle by indicating the coordinates of the 3 vertices; if your answer is No, state your reason.

[One way to do this problem is to use the result in Problem 4; but there are other ways too.]

1991–1992 First Round (November 16, 1991)

1. A palindromic number is an integer that reads the same forward or backward; for example, 1991 is a palindromic number. What is the difference of the two palindromic numbers closest to 1991 (but not 1991 itself)?

2. Lines from the vertices of a parallelogram to the midpoints of the sides are drawn as shown in Figure A7 forming a smaller parallelogram of area 7 cm^2 in the center. What is the area of the original parallelogram?

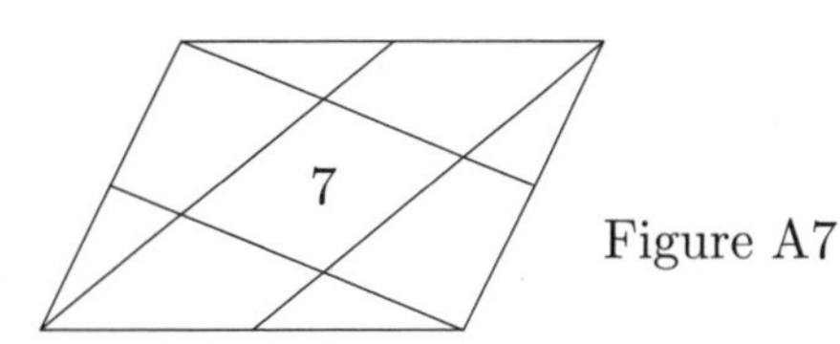

Figure A7

3. Let a, b, c and k be nonzero real numbers satisfying the relations

$$k = \frac{a}{b+c} = \frac{b}{c+a} = \frac{c}{a+b}.$$

Find all the possible common values k of these fractions.

4. Figure (a) and Figure (b) are examples of quadrangles whose four corners can be folded to meet at a point without either overlapping or forming a gap. Of the next 6 quadrangles (Figures (c) - (h)), determine which one(s) can be folded in the same manner.

Figure (c) $\overline{AB} = \overline{CD}$, $\overline{AD} = \overline{BC}$, $\overline{AB} \neq \overline{AD}$, $\angle A = 90^0$.

Figure (d) $\begin{cases} KN \perp KL. \\ K,\ L,\ M,\ N \text{ are midpoints of the respective sides.} \end{cases}$

Figure (e) $\begin{cases} KM \perp LN,\ \overline{KM} \neq \overline{LN}. \\ K,\ L,\ M,\ N \text{ are midpoints of the respective sides.} \end{cases}$

Figure (f) $\overline{AB} = \overline{AD}$, $\overline{BC} = \overline{CD}$.

Figure (g) $\overset{\frown}{AB} + \overset{\frown}{CD} = \overset{\frown}{BC} + \overset{\frown}{AD}$.

Figure (h) $AC \perp BD$.

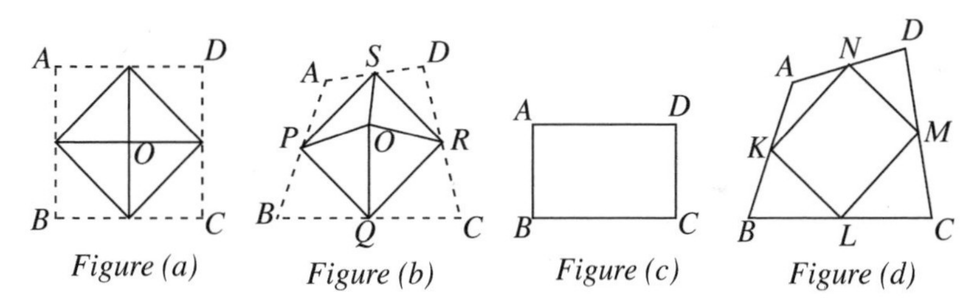

Figure A8

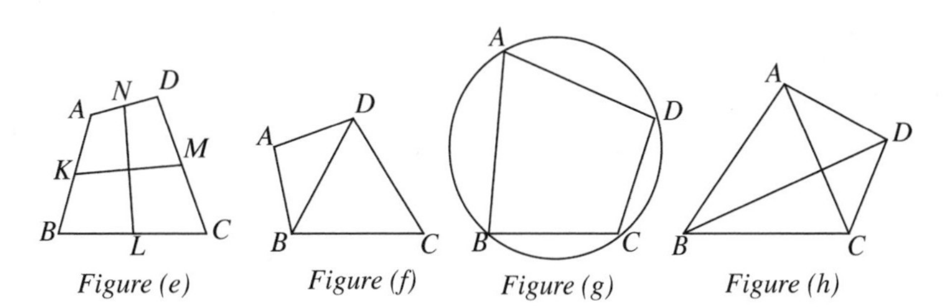

5. In a magic square of addition, the sum of the integers in each row, column, and diagonal is the same. For example, in Figure A9. the magic sum is 15. Fill in the blanks in Figure A10 with positive integers to make it a magic square of multiplication; i.e. complete Figure A10 with positive integers so that the product of the three positive integers in each row, column, and diagonal becomes the same.

   What is the sum of the two largest integers in the completed magic square of multiplication (Figure A10)?

4	9	2
3	5	7
8	1	6

Figure A9

3		
4		
	1	

Figure A10

6. In Figure A11, the three line segments AB, EF, CD are perpendicular to the line BD, and the three points A, E, D are collinear (i.e., on a straight line), as are the points B, E, C and B, F, D. Suppose $\overline{AB} = 10\ cm$, $\overline{CD} = 15\ cm$.

   Find the length $\overline{EF}$.

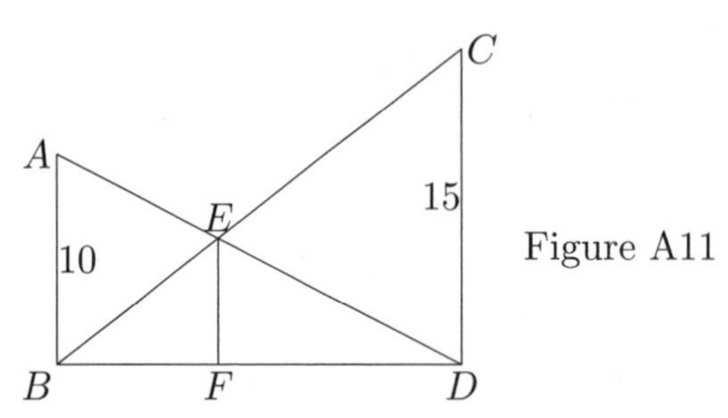

Figure A11

7. Suppose the two quadratic equations

$$x^2 - 5x + k = 0 \quad \text{and} \quad x^2 - 9x + 3k = 0$$

   have a nonzero root in common. What is the value of k?

8. Suppose the diagonals AC and BD of a trapezoid $ABCD$ intersect at P. If the areas of $\triangle PAB$ and $\triangle PCD$ are 36 cm^2 and 25 cm^2, respectively, what is the area of the trapezoid $ABCD$? (Note that $AB \parallel CD$.)

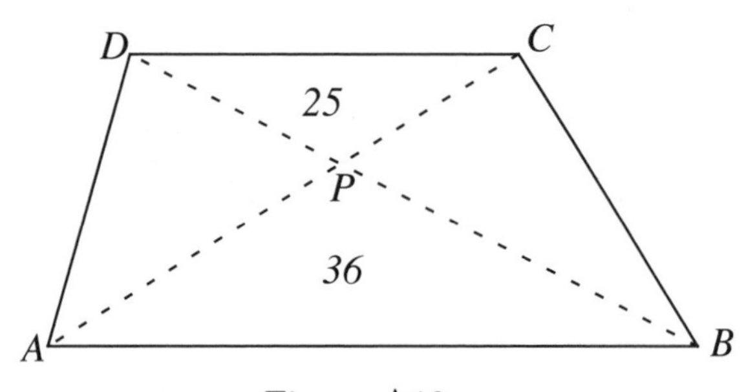

Figure A12

1991–1992 Final Round (February 1, 1992)

1. Figure A13 is an example of a magic square of order 4. The sum of the four integers in each row, column, or diagonal is always 34. So we say the magic sum of this magic square is 34. Suppose there is a magic square of order 7, with integers from 1 through 49. What would be the magic sum?

16	3	2	13
5	10	11	8
9	6	7	12
4	15	14	1

Figure A13

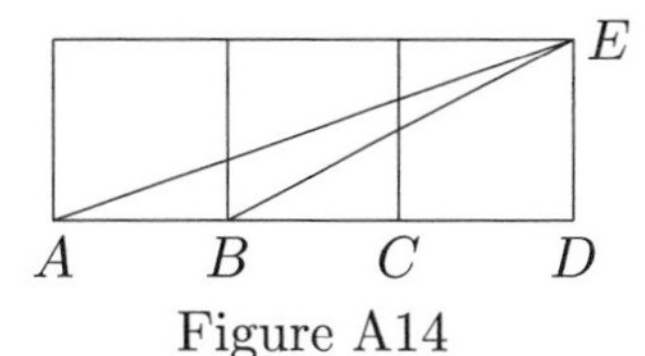

Figure A14

2. Arrange three squares of the same size as in Figure A14.

   Find the sum of the two angles $\angle DAE$ and $\angle DBE$.

3. Suppose a, b, c are the lengths of the sides of a triangle. If $a^2 + ab + b^2 = c^2$, then the largest angle of the triangle is $\frac{2\pi}{3}$. Observe that

$$\begin{array}{lllllllll} 5^2 & + & 5 & \times & 3 & + & 3^2 & = & 7^2, \\ 7^2 & + & 7 & \times & 8 & + & 8^2 & = & 13^2, \\ 9^2 & + & 9 & \times & 15 & + & 15^2 & = & 21^2, \\ 11^2 & + & 11 & \times & 24 & + & 24^2 & = & 31^2, \\ 13^2 & + & 13 & \times & 35 & + & 35^2 & = & 43^2. \end{array}$$

   Find positive integers x and y such that $17^2 + 17x + x^2 = y^2$.

4. In $\triangle ABC$,

$$\overline{BC} = 40\ cm, \quad \overline{CA} = 24\ cm, \quad \angle C = \frac{2\pi}{3}.$$

   Suppose the bisector of $\angle C$ meets the opposite side AB at D.

   (a) Find the length $\overline{AB}$.

   (b) Find the length $\overline{AD}$.

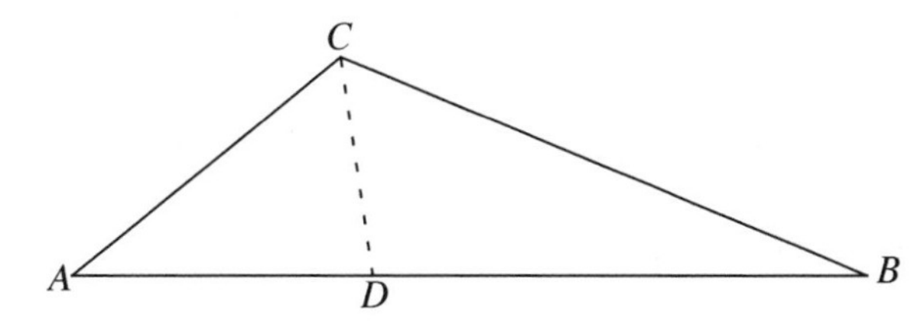

Figure A15

5. On reviewing last year's New Mexico Mathematics Contest problems, Debbie said: "A perfect square whose last digit is 6 must have an odd digit right before the last digit." Irene said: "I was thinking the converse: a perfect square whose next-to-the-last digit is odd, must have 6 as the last digit."

   Prove or disprove each of their assertions.

6. A square sheet of paper $ABCD$ is folded as shown in Figure A16 with D falling on E, which is on BC, with A falling on F, and EF intersecting AB at G.

   Prove that, regardless of the position of E on the side BC, the perimeter of $\triangle EBG$ is a fixed length, and express this length in terms of the length ℓ of a side of the square $ABCD$.

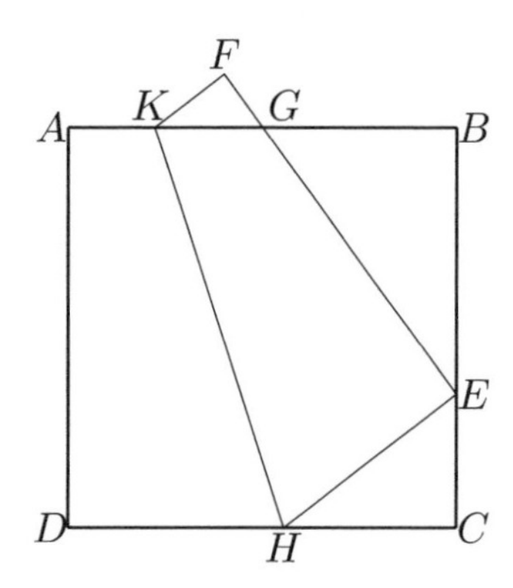

Figure A16

7. Suppose a cubic polynomial $x^3 + px^2 + qx + 72$ is divisible by both $x^2 + ax + b$ and $x^2 + bx + a$ (where a, b, p, q are constants and $a \neq b$).

   Find the zeros (roots) of the cubic polynomial.

8. Suppose the diagonals AC and BD of a convex quadrangle $ABCD$ intersect at P, and the areas of $\triangle PAB$ and $\triangle PCD$ are 16 cm^2 and 25 cm^2, respectively.

   What is the minimum possible area of the quadrangle $ABCD$?

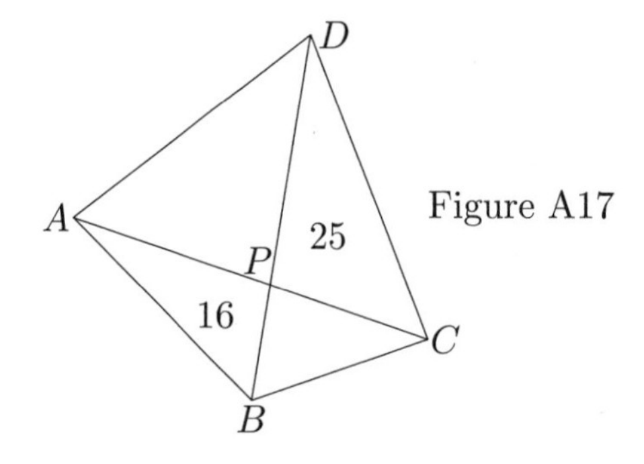

Figure A17

1992–1993 Final Round (November 14, 1992)

1. Evaluate
$$\frac{1992^3 - 1991 \times 1992 \times 1993}{2 \times 3 \times 83}.$$

2. In Figure A18, $ABCD$ is a rectangle and AEB is an isosceles right triangle (i.e., $\overline{AE} = \overline{BE}$ and $\angle AEB = \frac{\pi}{2}$). Suppose
$$\overline{AB} = 4\ cm, \quad \overline{BC} = 3\ cm.$$
Find the area of $\triangle AEC$.

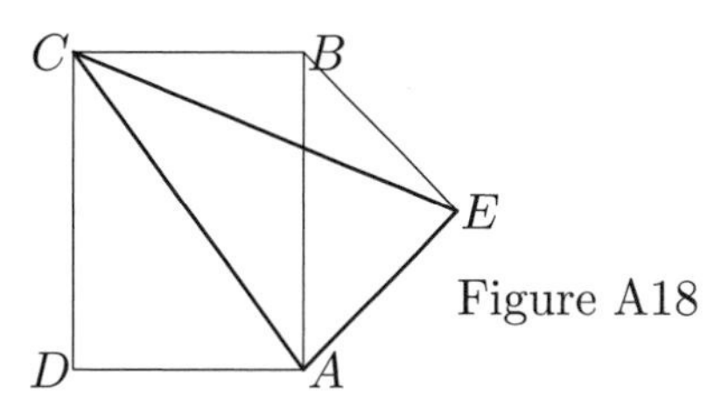

Figure A18

3. Two swimmers, Sam and Jim, at the opposite ends of a 50-meter pool, start simultaneously to swim the length of the pool; Sam swims at the rate of $\frac{3}{4}$ of a meter per second and Jim at $\frac{2}{3}$ of a meter per second. They swim back and forth with no loss of time at the turns.
   (a) How long will it take for Sam to catch up to Jim from behind?
   (b) How many times do Sam and Jim pass each other (swimming in the opposite direction) during this period (i.e., before Sam catches up to Jim for the first time)?

4. Suppose the incircle of $\triangle ABC$ touches the three sides BC, CA, AB at D, E, F as in Figure A19. If
$$\overline{AB} = 13\ cm, \quad \overline{BC} = 14\ cm, \quad \overline{CA} = 15\ cm,$$
then
   (a) what is the length $\overline{CE}$?
   (b) what is the radius r of the incircle?

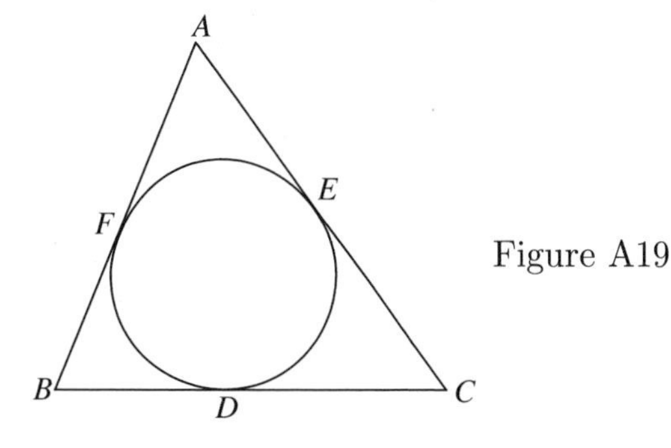

Figure A19

5. Find all real numbers x satisfying the inequality

$$\frac{x-1}{x-3} \geq \frac{x-2}{x-4}.$$

6. There are 30 lattice points, 5 by 6, as in Figure A20. Of all the squares with their 4 vertices at these lattice points,

   (a) how many have a pair of horizontal sides?

   (b) how many do not have sides that are horizontal?

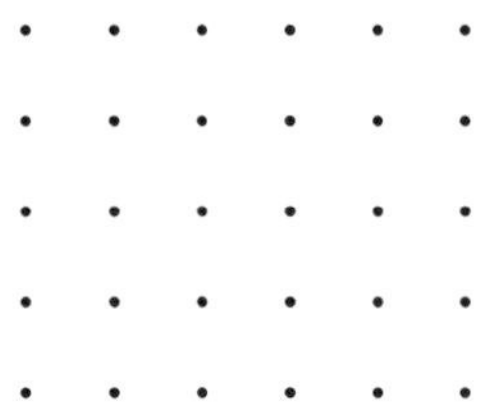

Figure A20

7. Let

$$\begin{aligned} f_1(x) &= x, & f_2(x) &= 1-x, & f_3(x) &= \frac{1}{x}, \\ f_4(x) &= \frac{1}{1-x}, & f_5(x) &= \frac{x}{x-1}, & f_6(x) &= \frac{x-1}{x}. \end{aligned}$$

   (a) Suppose $f_6\left(f_m(x)\right) = f_4(x)$. Then $m = ?$

   (b) Suppose $f_n\left(f_4(x)\right) = f_3(x)$. Then $n = ?$

8. Note that the set of all the points (x, y) in the plane satisfying the inequality

$$|x| + |y| \leq 1$$

is a square whose area is 2 (square units).

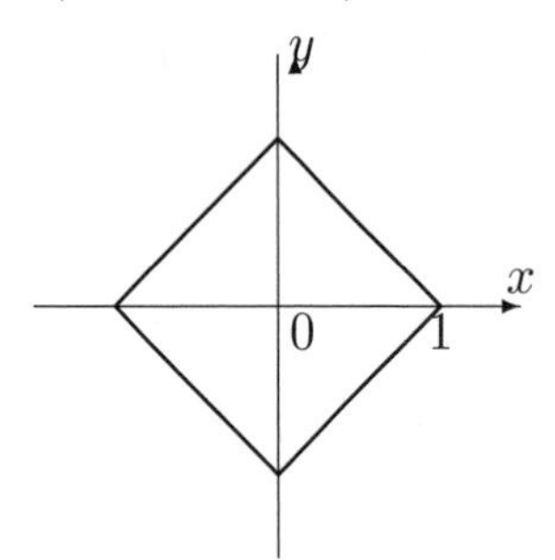

Figure A21

The set of all the points (x, y) in the plane satisfying the inequality

$$\Big||x|-1\Big| + \Big||y|-1\Big| \leq 2$$

also forms a polygon.

   (a) How many sides has this polygon?

   (b) What is its area?

1992–1993 Final Round (February 6, 1993)

1. (a) Find the remainder when 1993^2 is divided by 9.

   (b) For some integer exponent n, we have

$$1993^n = 15777A7325840B,$$

   where two missing digits are replaced by A and B. Find A, B and n.

2. (a) What is the maximum number of acute angles a convex polygon can have ? (A set is convex if it includes, with each pair of its points, the entire line segment joining them; e.g., every triangle is convex, and so is a circle, but neither a star nor a cross is convex. An angle is acute if it is less than $\frac{\pi}{2}$.)

   (b) If a convex polygon has exactly 5 obtuse angles, what is the maximum possible number of its sides? (An angle is obtuse if it is more than $\frac{\pi}{2}$.)

3. Figure A22 and Figure A23 are examples of magic squares of order 3 and 5, respectively. (In each case, the sum of the integers in each row, column, or diagonal is the same.)

19	93	32
61	48	35
64	3	77

Figure A22

8	21	17	5	14
2	15	9	23	16
24	18	1	12	10
11	7	25	19	3
20	4	13	6	22

Figure A23

Observe that the number 48 at the center of the magic square of order 3 is the average of all the entries in that magic square:

$$48 = \frac{1}{9}(19 + 93 + 32 + 61 + 48 + 35 + 64 + 3 + 77).$$

Is this always true of every magic square of order 3? That is, if Figure A24 is a magic square, then is it necessary that

$$E = \frac{1}{9}(A + B + C + D + E + F + G + H + I)?$$

A	B	C
D	E	F
G	H	I

Figure A24

Justify your assertion.

4. Let P be a point in the plane of an equilateral triangle ABC such that

$$\triangle PBC, \quad \triangle PCA, \quad \triangle PAB$$

are all isosceles. (A triangle is isosceles if at least two of its sides are of the same length; it is equilateral if all three of its sides are of the same length.)

How many such points P are there?

5. Find a function f satisfying the functional equation

$$f\left(\frac{1}{1-x}\right)+f\left(\frac{x-1}{x}\right)=\frac{x}{x-1} \quad \text{for all } x\neq 0,\ 1.$$

Is such a function unique?

6. Let ℓ_1, ℓ_2, ℓ_3 be three parallel lines with ℓ_2 between ℓ_1 and ℓ_3. Suppose the distance between ℓ_1 and ℓ_2 is a cm, and that of ℓ_2 and ℓ_3 is b cm.

   Express the area of an equilateral triangle having one vertex on each of the three parallel lines in terms of a and b.

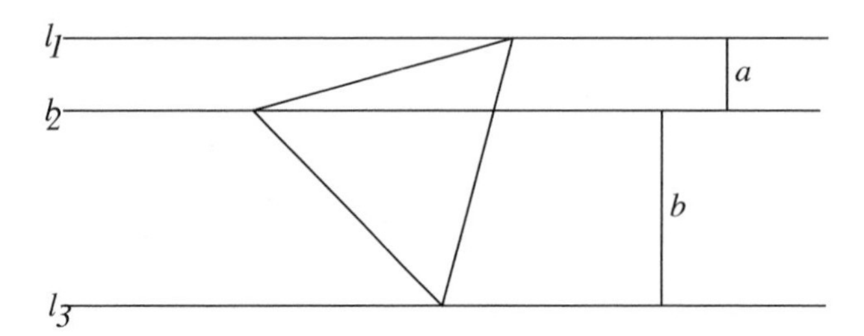

Figure A25

1993–1994 First Round (November 13, 1993)

1. Let a, b, c be integers satisfying

$$24a = 45b = c^2.$$

Find the smallest such positive integer c.

2. Let P be a point inside the square $ABCD$ such that $\triangle PAB$ is an equilateral triangle. Find $\angle CPD$.

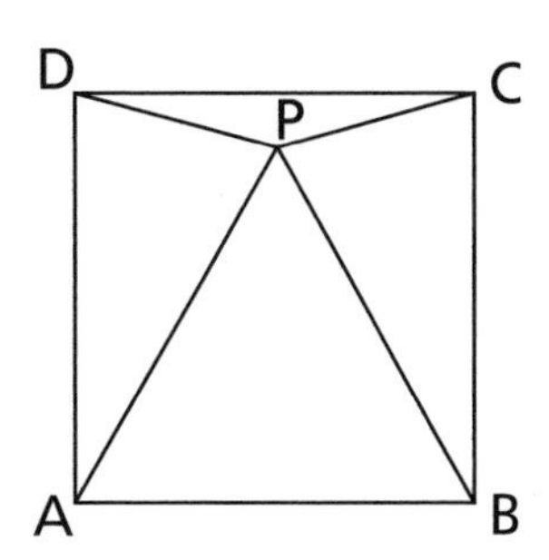

Figure A26

3. Hester has a pocketful of pennies, nickels, dimes,and quarters. (There is at least one coin of each value.)

If she has a total of 19 coins worth 93 cents, how many dimes does she have?

4. We have a box (a rectangular parallelepiped) whose height, width, and length are 3, 4, and 5 cm, respectively.

Suppose the box is sliced by a plane so that the cross-section is a square.

(a) What is the maximum possible length of a side of the square? How many distinct slices are there of this type?

(b) What is the minimum possible length of a side of the square? How many distinct slices are there of this type?

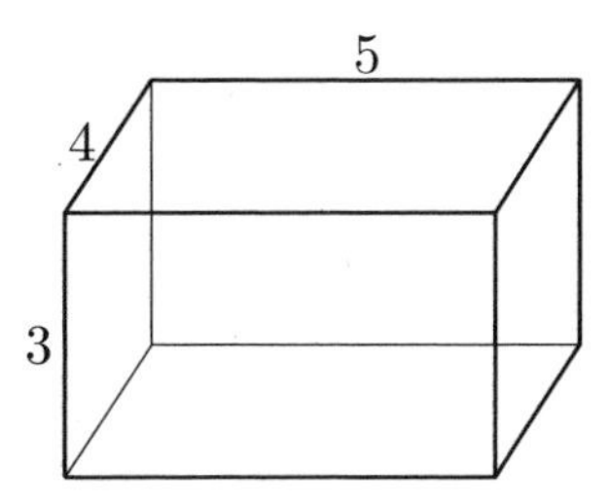

Figure A27

5. (a) Find positive integers u and v satisfying

$$\sqrt{18 - 2\sqrt{65}} = \sqrt{u} - \sqrt{v}.$$

(b) Find positive integers x and y satisfying

$$\sqrt{14+3\sqrt{3+2\sqrt{5-12\sqrt{3-2\sqrt{2}}}}} = x + \sqrt{y}.$$

6. Suppose a rectangular sheet of paper $ABCD$ is folded along the diagonal BD such that A falls on E, and F is the intersection of BE and CD. If

$$\overline{AB} = 4\ cm \quad \text{and} \quad \overline{AD} = 3\ cm,$$

then what is the area of $\triangle BDF$?

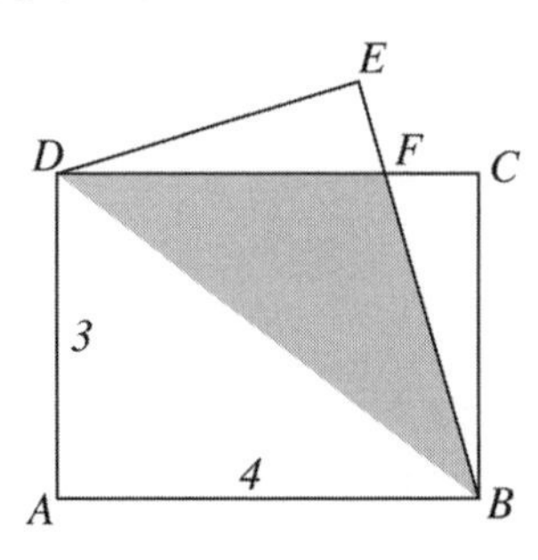

Figure A28

7. (a) Find the positive integers p and q (which have no common divisor other than 1) such that

$$\frac{p}{q} = 0.\dot{1}8\dot{5} = 0.185185185\cdots.$$

(b) Find the positive integers r and s (which have no common divisor other than 1) such that

$$\frac{r}{s} = 0.1\dot{4}8\dot{6} = 0.1486486486\cdots.$$

8. Points D and E are the respective midpoints of the sides AB and AC of a triangle ABC, and F is the intersection of the medians BE and CD. Suppose the area of the quadrangle $ADFE$ is 12 cm^2; what is the area of $\triangle ABC$?

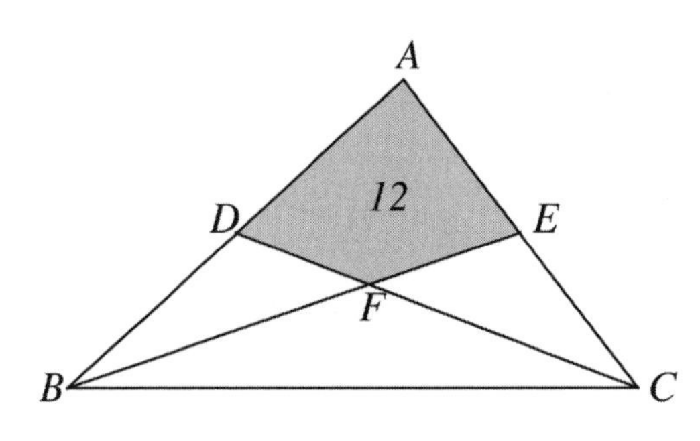

Figure A29

1993–1994 Final Round (February 5, 1994)

1. (a) Find the last digit in the decimal expression for the product of all the prime numbers less than 1994:

$$1993 \times 1987 \times 1979 \times \cdots \times 19 \times 17 \times 13 \times 11 \times 7 \times 5 \times 3 \times 2.$$

   (b) The factorial, $n!$, of a positive integer n is defined as the product of all the positive integers less than or equal to n. For example, $2! = 2 \times 1$, $3! = 3 \times 2 \times 1$, $7! = 7 \times 6 \times 5 \times 4 \times 3 \times 2 \times 1$, etc.

   Find the last digit in the decimal expression for the sum of all the factorials $p!$ where p runs through all the prime numbers less than 1994:

$$1993! + 1987! + 1979! + \cdots + 7! + 5! + 3! + 2!$$

2. A quadrangle has 2 diagonals while a pentagon has 5 diagonals. Suppose a polygon has 20 diagonals. How many sides does the polygon have?

3. Solve for x:

$$\frac{x+2}{x+1} + \frac{x+5}{x+4} = \frac{x+3}{x+2} + \frac{x+4}{x+3}.$$

4. A square sheet of paper $ABCD$ is folded along GH with A falling on E, which is on BC, and D falling on F. Suppose

$$\overline{AB} = 18\ cm, \quad \overline{BE} = 6\ cm.$$

   Find the area of the trapezoid $EFGH$.

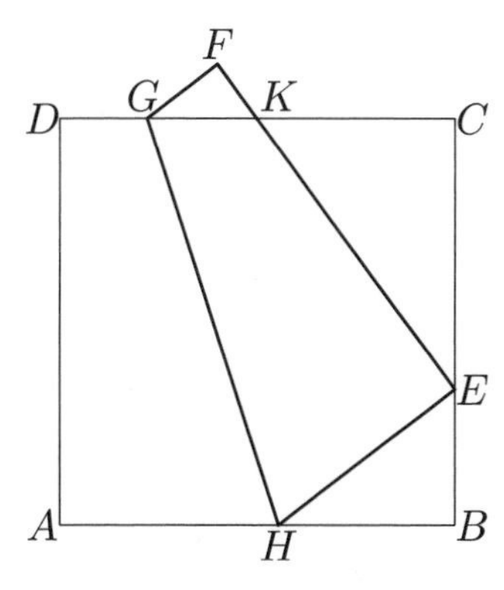

Figure A30

5. (a) Find all the possible denominators that give repeating decimals of minimum length 3 and without non-repeating parts; i.e., find all the positive integers q with the property that, given a positive integer p which has no common divisor (other than 1) with q, there exist 3 digits a, b, c (not all equal) and a non-negative integer n such that

$$\frac{p}{q} = n.\dot{a}b\dot{c} = n.abcabcabc\cdots.$$

(b) Fill in all the blanks, and show that the solution is unique.

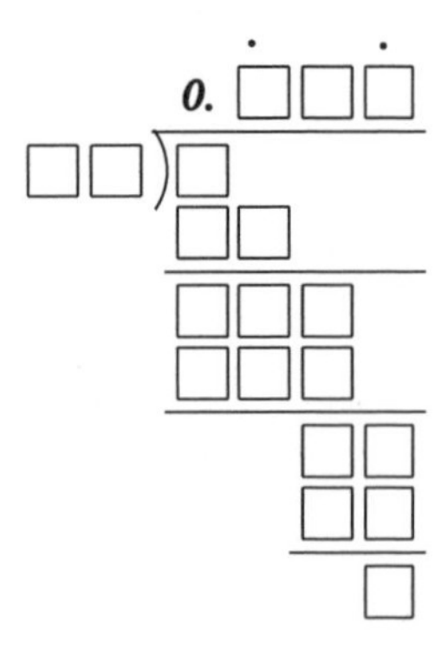

Figure A31

6. (a) A cube is sliced by a plane so that the cross-section is an equilateral triangle. What is the maximum possible length of a side of such an equilateral triangle (assuming the length of each edge of the cube is 1)? How many distinct slices are there of this type?

   (b) A cube is sliced by a plane so that the cross-section is a regular hexagon. What is the length of a side of such a regular hexagon (assuming the length of each edge of the cube is 1)? How many distinct slices are there of this type?

7. One summer day, Nicky visited Uncle Paul in Los Alamos.

   "Uncle Paul, show me some of your mathematics tricks," said Nicky.

   "O.K., if you wish," replied Uncle Paul. "Choose a number between 0 and 100, and just tell me the remainders when it is divided by 3, 5 and 7, respectively. Then I'll guess your secret number."

   "Let me see, my secret number gives the remainders 2, 4 and 5, when it is divided by 3, 5 and 7, respectively," said Nicky.

   "Abracadabra! Your secret number is 89," replied Uncle Paul within seconds.

   "Right! But how did you find it so fast?" asked Nicky.

   "It is very simple, Nicky," said Uncle Paul. "All you need is to remember the magic triple $\{70, 21, 15\}$. Your secret number gives the remainder 2 when divided by 3, so I multiply 70 by 2 to get 140. Similarly, I multiply 21 by 4 to get 84, and 15 by 5 to get 75. Then the sum $140 + 84 + 75 = 299$ has the property that it gives the same remainders as your secret number when it is divided by 3, 5 and 7. But 299 is not between 0 and 100, so I reduce it by a multiple of 105 $(= 3 \times 5 \times 7)$, to get 89 $(= 299 - 2 \times 105)$, which is your secret number."

   "That's neat, Uncle Paul. I'll think it over tonight. Thank you and good-bye."

   Early next morning, Nicky went to see Uncle Paul again.

   "Uncle Paul, choose a number between 0 and 1000, and just tell me the remainders when it is divided by 7, 11 and 13, respectively. Then I'll guess your secret number."

"O.K., my secret number gives remainders 1, 9 and 4 when it is divided by 7, 11 and 13, respectively."

Within a minute, Nicky was able to find Uncle Paul's secret number correctly.

(a) What is Uncle Paul's secret number?

(b) What is Nicky's magic triple?

8. Let P be a point inside an equilateral triangle ABC. The distances $\overline{PA}$, $\overline{PB}$, $\overline{PC}$ are 1, 2, $\sqrt{3}$ cm, respectively.

   (a) Find the angles around the point P; i.e., find $\angle BPC$, $\angle CPA$ and $\angle APB$.

   (b) Find the length of side BC of the equilateral triangle ABC.

1994–1995 First Round (November 12, 1994)

1. How many weeks is 10! seconds?
   (Note that $10! = 1 \times 2 \times 3 \times 4 \times 5 \times 6 \times 7 \times 8 \times 9 \times 10$.)

2. The circumcenter of a triangle is the center of the circle that passes through all three vertices of the triangle. Suppose O is the circumcenter of $\triangle ABC$, and $\angle OAB = 63^0$; find $\angle C$. Is the answer unique?

3. Let $y = x + \dfrac{1}{x}$. Then $x^2 + \dfrac{1}{x^2} = \left(x + \dfrac{1}{x}\right)^2 - 2 = y^2 - 2$.

   (a) Express $x^4 + \dfrac{1}{x^4}$ in terms of y.

   (b) Express $x^5 + \dfrac{1}{x^5}$ in terms of y.

4. Suppose there are eight points on a circle, equally spaced.

   (a) How many triangles are there having all their vertices at three of these eight points?

   (b) How many of these triangles are right triangles?

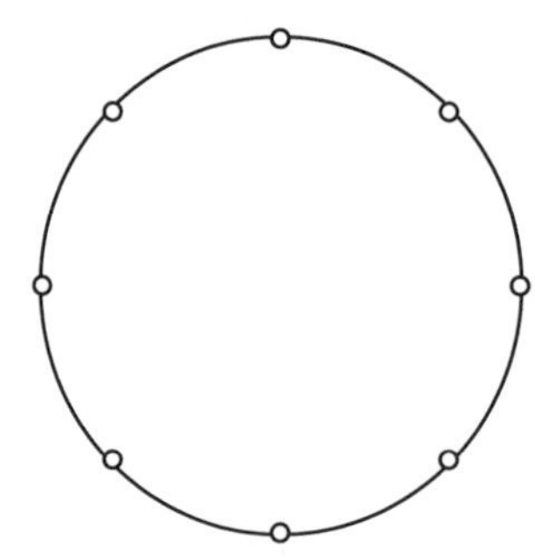

Figure A32

5. (a) Find the constants h and k such that the following becomes an identity

$$\frac{1}{x^2-1} = \frac{h}{x-1} + \frac{k}{x+1}.$$

   (b) Evaluate

$$\frac{1}{2^2-1} + \frac{1}{3^2-1} + \frac{1}{4^2-1} + \cdots + \frac{1}{10^2-1}.$$

   Express your answer as a fraction in lowest terms.

6. Suppose A, B, C, D are cocyclic (i.e., these four points are on a circle), K is the intersection of (the extensions of) the chords AB and CD, and

$$\overline{KA} = 4, \quad \overline{KB} = 3, \quad \overline{KC} = 2.$$

   (a) Assuming that K is inside the circle, find the length $\overline{KD}$.

   (b) Assuming that K is outside the circle, find the length $\overline{KD}$.

7. Find the positive integers a, b, c and p satisfying

$$\sqrt{94 + a\sqrt{p}} = b + c\sqrt{p},$$

where p is a prime number and $c > 1$.

8. Suppose AB is the common chord of two circles O and O', C and D are points on the circle O and O', respectively, such that BC is tangent to the circle O', while BD is tangent to the circle O. Suppose

$$\overline{AC} = 3, \quad \overline{AD} = 4;$$

find the length of the common chord AB.

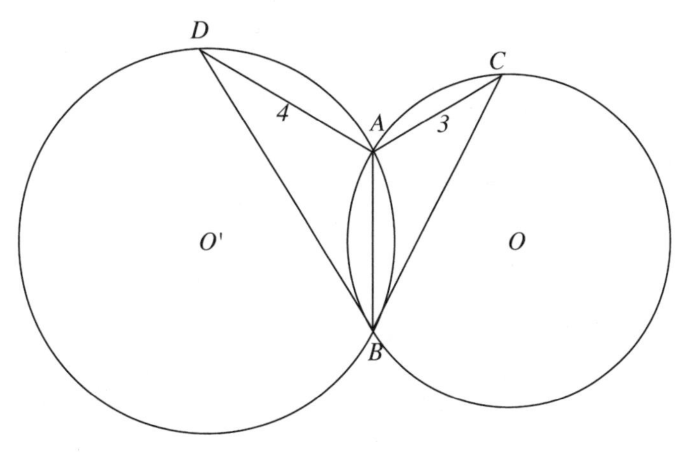

Figure A33

1994–1995 Final Round (February 4, 1995)

1. A rectangular chocolate bar consists of 30 small rectangular chocolate pieces arranged 5 by 6. If you "fold" it along a "valley", then you will break it into two pieces. What is the minimum number of "foldings" needed to break the chocolate bar into 30 small pieces? You are not allowed to arrange two or more pieces together in one "folding"; i.e., for each "folding", you are only permitted to break one piece into two.

2. Suppose there are 12 points on a circle, equally spaced.

   (a) Of all the triangles having all their vertices at three of these 12 points, how many have at least one 60^0 angle?

   (b) How many of these triangles are of the type 30^0-60^0-90^0?

3. (a) Find all possible values of y, where $y = x + \frac{1}{x}$, and x satisfies the equation

$$x^4 + 2x^3 - 22x^2 + 2x + 1 = 0.$$

   (b) Solve the quartic polynomial

$$x^4 + 2x^3 - 22x^2 + 2x + 1 = 0.$$

4. In Figure A34, points A, B, G, S are on a circle whose center is at M, and O is a point on the extension of the diameter AB. Moreover, OG is tangent to the circle and both GH and SM are perpendicular to the diameter AB. Suppose $\overline{OA} = a$ and $\overline{OB} = b$ $(0 < a < b)$. Express the inequalities

$$\overline{OH} < \overline{OG} < \overline{OM} < \overline{OS}$$

in terms of a and b.

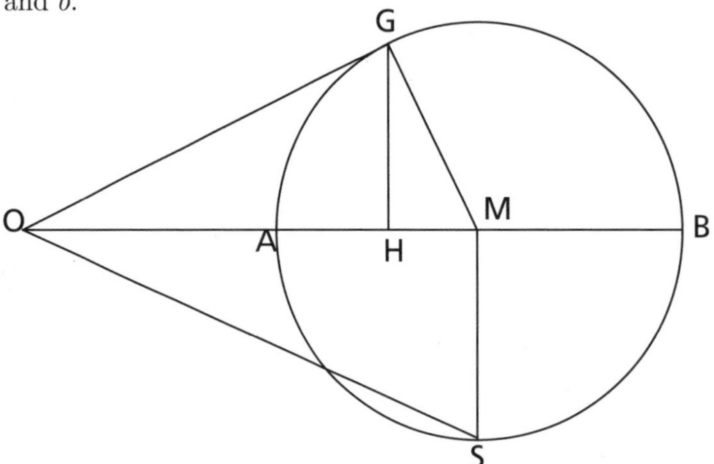

Figure A34

5. Suppose $a_1, a_2, \cdots, a_n, \cdots$ is a sequence of numbers with the property that

$$a_1 + a_2 + \cdots + a_n = \frac{n(n+1)(n+2)}{6} \quad \text{for all } n = 1, 2, 3, \cdots.$$

   (a) Find a_{19}.

   (b) Evaluate

$$\frac{1}{a_1} + \frac{1}{a_2} + \cdots + \frac{1}{a_{94}} + \frac{1}{a_{95}}.$$

6. In Figure A35, the side AB of a cyclic quadrangle $ABCD$ is a diameter of the circumcircle, and

$$\overline{BC} = 7, \quad \overline{CD} = \overline{DA} = 3.$$

What is the length of the diameter AB?

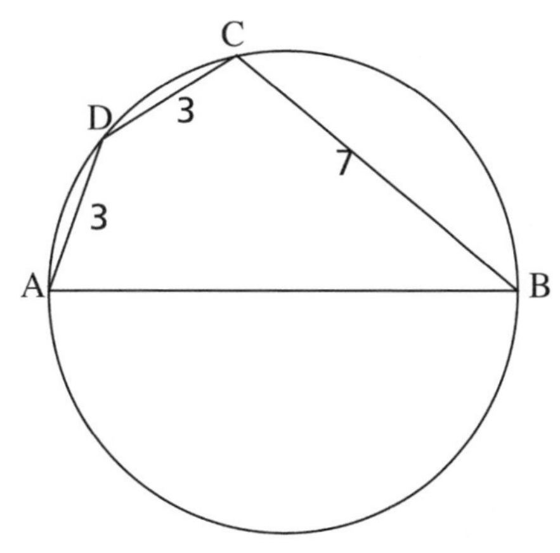

Figure A35

7. During the holidays, Sierra visited Aunt Lisa in California.
"Aunt Lisa, show me some of your mathematics tricks," said Sierra.
"O.K., if you wish," replied Aunt Lisa. "Choose any positive integer, say 314159, and rearrange its digits in any way you want and get, say 193415; then take (the absolute value of) their difference, in our case we have 314159 − 193415 = 120744. Now, you hide away one of the nonzero digits in the difference, and just tell me the rest of the digits in any order, say 4, 2, 1, 0, 4. Then I'll guess the nonzero digit you hid away; in our example, it's 7."
"Oh, that's a cool mathematics game to play with kindergarten kids, but I see the trick behind it right away," said Sierra.
"Really? O.K., let me test you," said Aunt Lisa with an air of doubt. "I choose a number, scramble its digits and find the digits in the difference to be 6, 0, 5, 8, 3, in some order, with one nonzero digit hidden away. Now tell me the nonzero digit I hid away."
Instantly, Sierra found Aunt Lisa's hidden digit correctly.

   (a) What is Aunt Lisa's hidden digit?

   (b) Explain the reason behind this trick.

8. Suppose P is a point inside a square $ABCD$ such that

$$\overline{PA} = 3, \quad \overline{PB} = 2\sqrt{2}, \quad \overline{PC} = 5.$$

(a) Find $\angle APB$.

(b) Find the length of side AB of the square $ABCD$.

1995–1996 First Round (November 11, 1995)

1. Observe that the greatest common divisor of the pair {18, 24} is 6, and the least common multiple is 72. The same is true for the triple {12, 18, 24} and the quadruple {12, 18, 24, 36}.

   (a) Find all pairs of positive integers whose greatest common divisor is 95, and whose least common multiple is 1995.

   (b) Find all triples of positive integers whose greatest common divisor is 95, and whose least common multiple is 1995.

   (c) Find all quadruples of positive integers whose greatest common divisor is 95, and whose least common multiple is 1995.

2. (a) Of all the triangles having all of their vertices at the vertices of a given cube, how many of them are right triangles?

   (b) Describe the remaining triangles, if any, that are not right triangles. How many triangles are there of this type?

3. Observe that

$$\begin{aligned} 3^2 + 4^2 &= 5^2, \\ 3^2 + 4^2 + 12^2 &= 13^2, \\ 3^2 + 4^2 + 12^2 + 84^2 &= 85^2. \end{aligned}$$

   Find a pair of positive integers x and y such that

$$3^2 + 4^2 + 12^2 + 84^2 + x^2 = y^2.$$

   (It is not necessary to find all such pairs.)

4. In a quadrangle $ABCD$, suppose

$$\overline{AB} = 5\ cm, \quad \overline{CD} = 8\ cm, \quad \angle ABC = 70^0, \quad \angle BCD = 50^0,$$

   and P, Q, R, S are the midpoints of BC, CA, AD, DB, respectively. Find the area of the parallelogram $PQRS$.

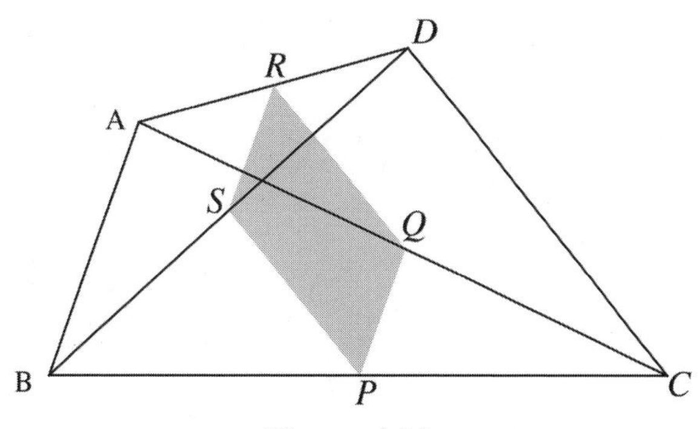

Figure A36

5. Suppose α and β are the two roots of the quadratic equation $3x^2 + 4x + 5 = 0$. Find the value of
$$\frac{\alpha}{\beta} + \frac{\beta}{\alpha}.$$
(Express your answer as a fraction in lowest terms.)

6. Suppose A and B are centers of the circles whose radii are 3 cm and 7 cm, respectively, and $\overline{AB} = 12\,cm$. Find the lengths of all common tangents PQ, where P and Q are on the circles A and B, respectively.

7. A function g is said to be even, while a function h is said to be odd, if
$$g(-x) = g(x), \qquad h(-x) = -h(x) \qquad \text{for all } x.$$
For example, $g(x) = 3 + 5x^2$ is even, while $h(x) = 2x - x^3$ is odd.

   (a) Given a function
   $$f(x) = \frac{1}{1 - x + x^2},$$
   find a pair of functions g and h, where g is even and h is odd, such that
   $$f(x) = g(x) + h(x) \quad \text{for all real numbers } x.$$

   (b) Is such a decomposition of f unique?

8. (a) Find integers a and b satisfying
   $$\tan\frac{\pi}{8} = \sqrt{a} - b.$$
   Hint: In the figure, $\angle ABC = \frac{\pi}{2}, \quad \overline{AB} = \overline{BC}, \quad \overline{AC} = \overline{CD}.$
   $$\therefore\ \angle ADB = \frac{\pi}{8}, \quad \tan\frac{\pi}{8} = \overline{AB}\big/\overline{BD}.$$

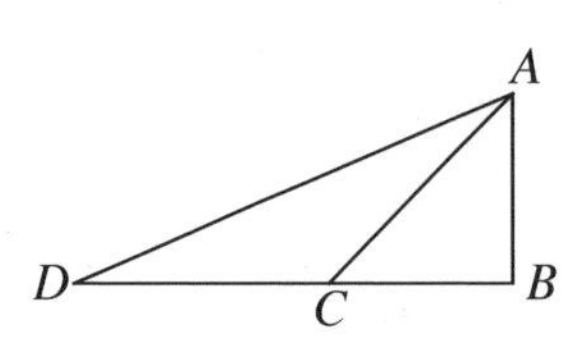

Figure A37

   (b) Find integers c and d satisfying
   $$\tan\frac{\pi}{12} = c - \sqrt{d}.$$

   (c) Find integers p, q, r, s satisfying
   $$\tan\frac{\pi}{24} = (\sqrt{p} - \sqrt{q})(\sqrt{r} - s).$$

1995–1996 Final Round (February 3, 1996)

1. In the freshman class at Euclid High School, there are 18 boys and 12 girls. The average height of the boys is 170 cm, and that of the girls is 160 cm. What is the average height of all the students in the class?

2. Suppose AC is one of the sides of an equilateral triangle having all of its vertices at the vertices of the cube obtained by folding the figure on the left. Indicate the other two sides (using the labels in the figure on the left). Is the answer unique?

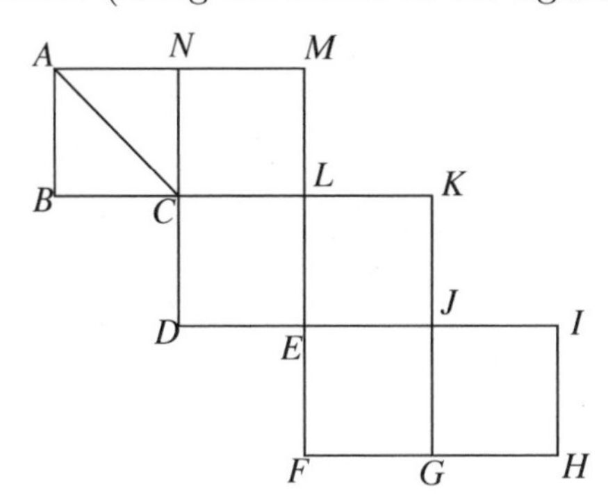

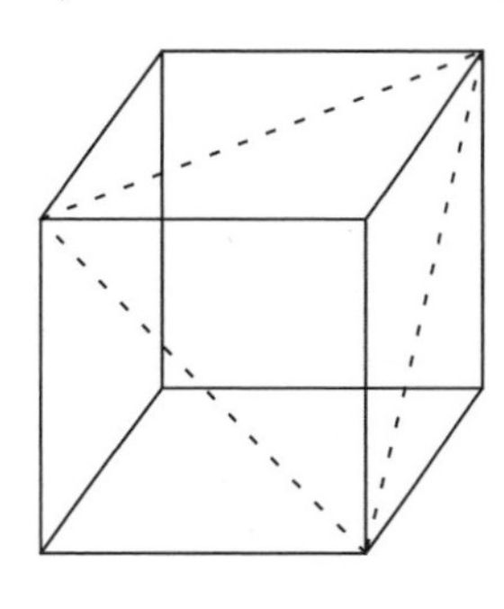

Figure A38

3. Find the minimum value of all the numbers in the sequence

$$\sqrt{\frac{7}{6}}+\sqrt{\frac{96}{7}},\quad \sqrt{\frac{8}{6}}+\sqrt{\frac{96}{8}},\quad \sqrt{\frac{9}{6}}+\sqrt{\frac{96}{9}},\ \cdots,\ \sqrt{\frac{95}{6}}+\sqrt{\frac{96}{95}}.$$

4. Of all the triangles having all their vertices at three of the vertices of a given convex n-gon (i.e., a convex polygon with n sides), $7n$ of them share no side with the polygon. Find the value of n.

5. Let α, β, γ be the three roots of the cubic equation $x^3 - 2x + 3 = 0$.

   (a) Find the value of $\alpha^2 + \beta^2 + \gamma^2$.

   (b) Find the value of $\alpha^3 + \beta^3 + \gamma^3$.

6. On the sides AB, AC of $\triangle ABC$, choose two points D and E, respectively, in such a way that $\triangle ADE$ and the quadrangle $DBCE$ have the same areas and the same perimeters. Given

$$\overline{BC} = 13\,cm,\quad \overline{CA} = 15\,cm,\quad \overline{AB} = 16\,cm,$$

find the lengths $\overline{AD}$ and $\overline{AE}$. Is the answer unique?

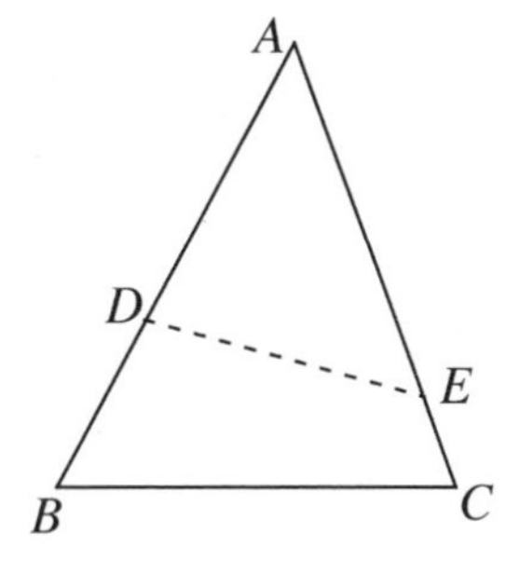

Figure A39

7. Observe that if

$$g(x) = x + \frac{1}{x}, \quad h(x) = \frac{x+2}{2x+1},$$

then, for all $x > 0$, we have

$$g\left(\frac{1}{x}\right) = g(x) > 0, \quad h\left(\frac{1}{x}\right) = \frac{1}{h(x)}.$$

(a) Given a function $f(x) = \dfrac{x}{x^2+3}$, find a pair of functions, g and h, where, for all $x > 0$, we have

$$g\left(\frac{1}{x}\right) = g(x) > 0, \quad h\left(\frac{1}{x}\right) = \frac{1}{h(x)} \quad \text{and} \quad f(x) = g(x) \cdot h(x).$$

(b) Is such a decomposition of f unique?

8. (a) Find positive integers a and b satisfying

$$\tan\frac{3\pi}{8} = a + \sqrt{b}.$$

(b) Show that

$$\left(\tan\frac{3\pi}{8}\right)^n + (-1)^n\left(\cot\frac{3\pi}{8}\right)^n$$

is an even integer for every positive integer n.

(c) For each positive integer n, let k_n be a positive integer such that

$$k_n < \left(\tan\frac{3\pi}{8}\right)^n < k_n + 1.$$

Show that k_n and n are of opposite parity; i.e., one is even and the other is odd.

(d) Find k_5.

The contest was suspended for the academic year 1996–1997

1997–1998 First Round (November 15, 1997)

1. Find positive integers x, y, and z satisfying

$$x + \frac{y}{19} + \frac{z}{97} = \frac{1997}{19 \times 97}.$$

2. Suppose a square is projected orthogonally to a line (the square and the line are coplanar) such that the images of the two diagonals have lengths 7 and 3 cm, respectively; i.e.,

$$\overline{A'C'} = 7, \quad \overline{B'D'} = 3,$$

in the figure. What is the length of a side of the square?

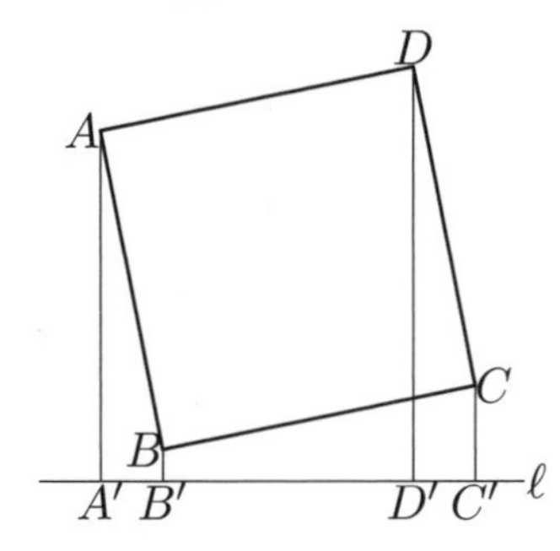

Figure A40

3. Assume there exist coefficients p, q, r such that, for every positive integer n,

$$\frac{1^4 + 2^4 + 3^4 + \cdots + (n-1)^4 + n^4}{1^2 + 2^2 + 3^2 + \cdots + (n-1)^2 + n^2} = pn^2 + qn + r.$$

Determine the coefficients p, q, r.

4. Suppose $\triangle AEF$ is inscribed in a rectangle $ABCD$ as in the figure. If the area of $\triangle AEF$ is 25 cm^2, and

$$\overline{BE} = 4\ cm,$$
$$\overline{DF} = 6\ cm,$$

find the area of the rectangle $ABCD$.

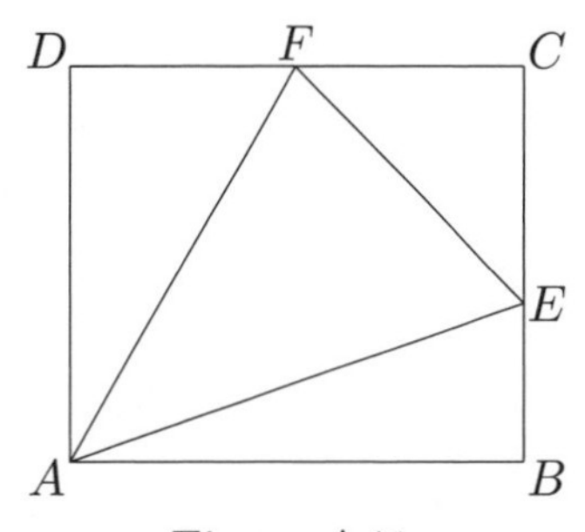

Figure A41

5. (a) Find integers a and b such that $\dfrac{4+3\sqrt{3}}{2+\sqrt{3}}$ is a root (zero) of the quadratic polynomial $x^2 + ax + b$.

   (b) Suppose

$$f(x) = x^4 + 2x^3 - 10x^2 + 4x - 10,$$

   and

$$f\left(\frac{4+3\sqrt{3}}{2+\sqrt{3}}\right) = c\sqrt{3} + d,$$

   where c and d are integers. Find the values of c and d.

6. We want to paint all five regions in the figure such that no neighboring regions are painted by the same color.

   (a) If we have four different colors, how many ways are there to paint the regions?

   (b) What if we have five different colors?

   It is not necessary to use all the colors.

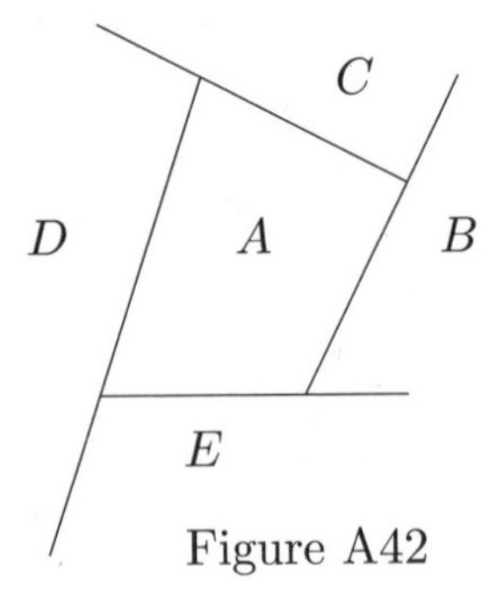

Figure A42

7. Suppose $u + v = 3,\ u^2 + v^2 = 13$.

   (a) Find the value of uv.

   (b) Find the value of $u^3 + v^3$.

8. In $\triangle ABC$, suppose

$$\overline{AB} = 5\,cm, \quad \overline{AC} = 7\,cm, \quad \angle ABC = \frac{\pi}{3}.$$

   (a) Find the length of the side BC.

   (b) Find the area of $\triangle ABC$.

1997–1998 Final Round (February 7, 1998)

1. Two quadratic equations

$$1997x^2 + 1998x + 1 = 0 \quad \text{and} \quad x^2 + 1998x + 1997 = 0$$

have a root in common. Find the product of the roots that are not in common.

2. Suppose a quadrangle $EFGH$ is inscribed in a rectangle $ABCD$ as in the figure. Let G' and H' be points on AB and BC, respectively, such that GG' is parallel to DA, and HH' is parallel to AB. If the area of the quadrangle $EFGH$ is 32 cm^2, and

$$\begin{aligned} \overline{EG'} &= 3\,cm, \\ \overline{FH'} &= 5\,cm, \end{aligned}$$

find the area of the rectangle $ABCD$.

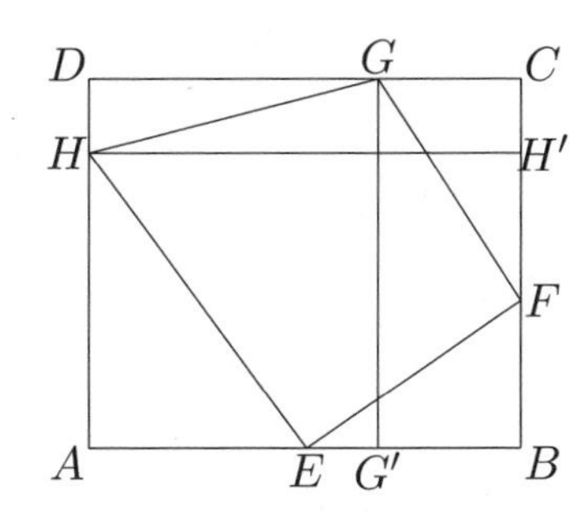

Figure A43

3. Find rational numbers x, y, u, v satisfying

$$\begin{aligned} z &= x - \sqrt{2}, & w &= 5 + y\sqrt{2}, \\ z + w &= u + \sqrt{2}, & zw &= 11 + v\sqrt{2}. \end{aligned}$$

4. Suppose $\sin\theta - \cos\theta = \dfrac{\sqrt{5}}{2}$.

   (a) Find the value of $\sin\theta \cdot \cos\theta$.

   (b) Find the value of $\sin^3\theta - \cos^3\theta$.

5. Find the positive integer k satisfying

$$\frac{1}{k+1} < \left(\sqrt{3} - \sqrt{2}\right)^4 < \frac{1}{k}.$$

6. Suppose a rectangle is projected orthogonally to a line (the rectangle and the line are coplanar) such that the images of the two diagonals have lengths 7 and 5 cm, respectively.

   (a) What is the minimum possible area of the rectangle?

   (b) Find the dimensions of the rectangle that gives the minimum area.

7. Observe that, for $k = 40$,

$$2k + 1 = 2 \cdot 40 + 1 = 9^2, \quad 3k + 1 = 3 \cdot 40 + 1 = 11^2.$$

We want to show that there exist infinitely many positive integers k for which $2k + 1$ and $3k + 1$ are both perfect squares. Suppose

$$2k + 1 = m^2, \quad 3k + 1 = n^2,$$

where m and n are positive integers. Define an integer K by the magic formula:

$$K = 11m^2 + 20mn + 9n^2.$$

Then $2K+1$ and $3K+1$ are both perfect squares, because for suitable positive integers p, q, r, s (and m, n as above),

$$2K + 1 = (pm + qn)^2, \quad 3K + 1 = (rm + sn)^2.$$

It follows that for any positive integer k for which $2k + 1$ and $3k + 1$ are both perfect squares, our magic formula gives a bigger integer K with the same property. Hence there exist infinitely many positive integers with the desired property. Determine the coefficients p, q, r, s.

8. (a) How many ways can the faces of a cube be painted by six different colors if no two faces are to be painted by the same color?

   (b) What if the cube in Part (a) is replaced by a box (i.e., a rectangular parallelopiped) none of whose faces is a square?

1998–1999 First Round (November 14, 1998)

1. Simplify

$$\frac{1}{2+\sqrt{5}}+\frac{1}{\sqrt{5}+\sqrt{6}}+\frac{1}{\sqrt{6}+\sqrt{7}}+\frac{1}{\sqrt{7}+\sqrt{8}}+\frac{1}{\sqrt{8}+3}.$$

2. We have several points in a plane; no two of them are less than 1 cm apart (i.e., every pair of points is at least 1 cm apart). What is the maximun possible number of points inside or on a circle of radius 1 cm?

3. Let $x=\dfrac{\sqrt{3}+\sqrt{2}}{\sqrt{3}-\sqrt{2}}, \quad y=\dfrac{\sqrt{3}-\sqrt{2}}{\sqrt{3}+\sqrt{2}}.$

   (a) Evaluate x^2+y^2.

   (b) Evaluate x^3+y^3.

4. Suppose E and F are the midpoints of sides BC and CD, respectively, of parallelogram $ABCD$. If the area of $\triangle AEF$ is 12 cm^2, what is the area of the parallelogram $ABCD$?

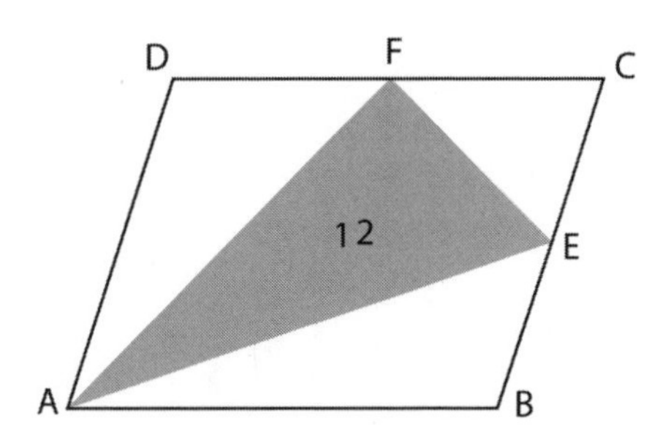

Figure A44

5. Expressions

$$s_1=a+b+c, \quad s_2=bc+ca+ab, \quad s_3=abc$$

are known as the basic symmetric forms of a, b, c. Every symmetric form of a, b, c can be expressed in terms of s_1, s_2, s_3. For example,

$$a^2+b^2+c^2=s_1^2-2s_2.$$

   (a) Express $(b+c)(c+a)(a+b)$ in terms of s_1, s_2, s_3.

   (b) Express $a^3+b^3+c^3$ in terms of s_1, s_2, s_3.

6. The circumradius of a triangle is the radius of the circle, called the circumcircle, that passes through the three vertices of the triangle. The inradius of a triangle is the

radius of the circle, called the incircle, that is inside the triangle and is tangent to the three sides of the triangle.

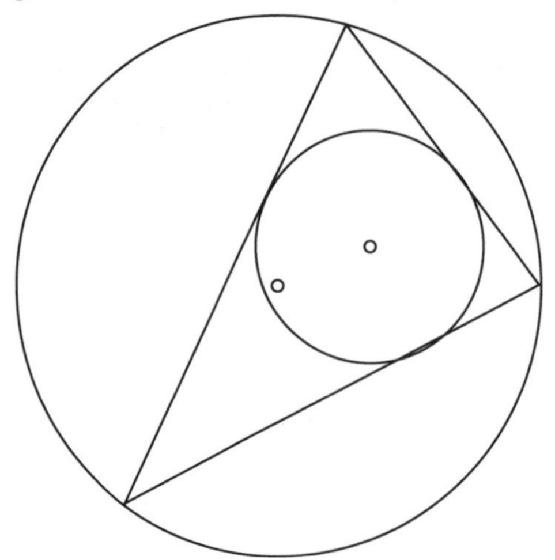

Figure A45

(a) Find the circumradius R of the 3-4-5 right triangle.

(b) Find the inradius r of the 3-4-5 right triangle.

7. (a) Find, if any, a function f satisfying

$$f\left(\frac{x}{x-1}\right) = x \quad (x \neq 1).$$

(b) Find, if any, a nonconstant function g satisfying

$$g\left(\frac{x}{x-1}\right) = g(x) \quad (x \neq 1).$$

How many such functions are there?

8. In $\triangle ABC$, let X and Y be the feet of perpendiculars from vertex A to the bisectors of (interior) angles B and C, respectively. Suppose the lengths of AB, AC and XY are 10, 7 and 3 cm, respectively. Find the length of side BC.

1998–1999 Final Round (February 6, 1999)

1. Find all prime numbers p such that $1999p + 1$ is a perfect square.

2. Lattice points are points (in the coordinate plane) whose coordinates are integers.

   (a) Does there exist a triangle similar to the 3-4-5 right triangle having all its vertices at lattice points, and having exactly one of its sides parallel to the coordinate axes?

   (b) Does there exist a triangle similar to the 3-4-5 right triangle having all its vertices at lattice points, yet having none of its sides parallel to the coordinate axes?

3. Two cousins, Amy and Kira, enjoy playing mathematics games together. One day, Amy said to Kira: "Choose any two numbers; then let the third number be the sum of the first and the second, the fourth number the sum of the second and the third, and so on. That is, from the third number on, each one is the sum of the two numbers immediately preceding it. For example, suppose your first two numbers are 2 and 1, respectively. Then you have the sequence:

$$2,\ 1,\ 3,\ 4,\ 7,\ 11,\ 18,\ 29,\ 47,\ 76,\ \cdots.$$

   Now just tell me the seventh number in your sequence, and I'll guess the sum of the first ten numbers in your sequence." "O.K.," said Kira. "Now I choose the first two numbers, and I find the seventh number in my sequence to be 99." Instantly, Amy found the sum of the first ten numbers in Kira's sequence correctly. What was Amy's answer?

4. (a) Show that there exists an angle φ such that

$$3\sin\theta + 4\cos\theta = 5\sin(\theta + \varphi) \quad \text{for all } \theta.$$

   (b) Find a necessary and sufficient condition on the constant k such that the trigonometric equation

$$k\sin\theta - 2\cos\theta = \sqrt{7}$$

   has a solution.

5. Let α and β be the roots of the quadratic equation

$$(x-2)(x-3) + (x-3)(x+1) + (x+1)(x-2) = 0.$$

   Evaluate

$$\frac{1}{(\alpha+1)(\beta+1)} + \frac{1}{(\alpha-2)(\beta-2)} + \frac{1}{(\alpha-3)(\beta-3)}.$$

6. Suppose square $DEFG$ is inscribed in $\triangle ABC$, with vertices D, E on side AB, and F, G on sides BC, CA, respectively. Given $\overline{AB} = 28\,cm$ and the length of a side of

square $DEFG$ is $12\,cm$, determine, if possible, the area of $\triangle ABC$.

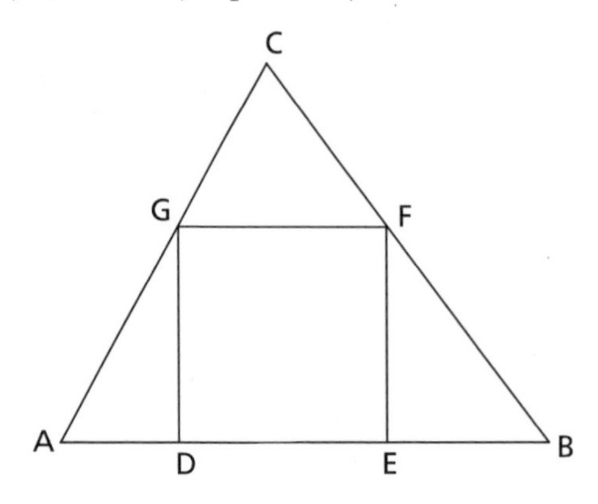

Figure A46

7. Observe that function

$$f(x) = x + \frac{1}{x}$$

has the properties that

$$\begin{aligned} f\left(\frac{1}{x}\right) &= f(x) \quad \text{for } x > 0; \quad \text{and} \\ f(x) - f(1) &= x + \frac{1}{x} - 2 = \frac{(x-1)^2}{x} \end{aligned}$$

has a double root at $x = 1$.

Suppose g is a nonconstant rational function such that

$$g\left(\frac{1}{x}\right) = g(x) \quad \text{for } x > 0.$$

State and prove a proposition about the multiplicity (order) of the root of

$$g(x) - g(1) = 0$$

at $x = 1$.

8. The circumradius of a triangle is the radius of the circle, called the circumcircle, that passes through the three vertices of the triangle. The inradius of a triangle is the radius of the circle, called the incircle, that is inside the triangle and is tangent to the three sides of the triangle.

   (a) Express the area of a right triangle in terms of its circumradius and inradius.

   (b) Express the lengths of the three sides of a right triangle in terms of its circumradius and inradius.

APPENDIX B

Answers to New Mexico Mathematics Contests

1990–1991 First Round (November 10, 1990)

1. 8 buses are needed.
2. $\overline{AB} = 17\ cm$.
3. November 7.
4. The sum of the seven angles is $3\pi\ (= 540°)$.
5. Esther will earn \$2700 more in the first 5 years.
6. The lengths of the two legs are 20 and 21 cm.
7. $d = 7$.
8. The only possible perfect square is (f) 5729581636.

1990–1991 Final Round (February 2, 1991)

1. 24 poles must be removed.
2. The area of the quadrangle is 58 cm^2.
3. There are 15 children and 170 candies.
4. $a = 1$, $b = \frac{1}{2}$, $c = -1$.
5. No, there exists no such perfect square.
6. The area of $\triangle ABC$ is 18 cm^2.
7. $n = 20$.
8. No, there exists no lattice equilateral triangle.

1990–1991 First Round (November 10, 1990)

1. The difference is 121.
2. The area of the parallelogram is 35 cm^2.
3. The possible values of k are $\frac{1}{2}$ and -1.
4. The figures which can be so folded are the ones labelled (d), (f), (g), (h).
5. The sum of the two largest integers is 54.
6. $\overline{EF} = 6\ cm$.
7. $k = 6$.
8. $[ABCD] = 121\ cm^2$.

1991–1992 Final Round (February 1, 1992)

1. The magic sum is 175.
2. The sum of the two angles is $\frac{\pi}{4}$ $(= 45°)$.
3. $x = 63, \qquad y = 73$.
4. $\overline{AB} = 56\ cm, \quad \overline{AD} = 21\ cm$.
5. Both of their assertions are correct.
6. The perimeter of $\triangle EBG$ is 2ℓ.
7. The roots of the given cubic polynomials are 1, 8 and -9.
8. The minimum possible area is 81 cm^2.

1992–1993 First Round (November 14, 1992)

1. 4.
2. 7 cm^2.
3. (a) 10 (minutes). (b) 8 times.
4. (a) $\overline{CE} = 8\ cm$. (b) $r = 4\ cm$.
5. $3 < x < 4$.
6. (a) 40. (b) 30.
7. (a) $m = 6$. (b) $n = 5$.
8. (a) 12-gon (dodecagon). (b) 24 (square units).

1992–1993 Final Round (February 6, 1993)

1. (a) The remainder is 7. (b) $A = B = 1$, $n = 4$.

2. (a) A convex polygon can have at most 3 acute angles. (b) 8 sides.

3. Yes, E is the average of all the entries.

4. There are ten such points.

5. $$f(x) = \frac{1}{2}\left\{(1-x) + \frac{1}{x} - \frac{x}{x-1}\right\} = \frac{1-x^2+x^3}{2x(1-x)}$$
   is the unique function satisfying the given functional equation.

6. $$\text{Area } = \frac{1}{\sqrt{3}}\{a^2 + ab + b^2\}.$$

1993–1994 First Round (November 13, 1993)

1. $c = 60$.

2. $\angle CPD = \frac{5\pi}{6}$.

3. Hester has two dimes.

4. (a) 5 cm; 2 ways. (b) 3 cm; infinitely many ways.

5. (a) $u = 13$, $v = 5$. (b) $x = 3$, $y = 2$.

6. $[BDF] = \dfrac{75}{16} = 4\dfrac{11}{16}\ cm^2$.

7. (a) $p = 5$, $q = 27$. (b) $r = 11$, $s = 74$.

8. $[ABC] = 36\ cm^2$.

1993–1994 Final Round (February 5, 1994)

1. (a) 0. (b) 8.

2. The polygon has eight sides (i.e., it is an octagon).

3. $x = \dfrac{5}{2}$.

4. $[EFGH] = 126\ cm^2$.

5. (a) 27, 37, 111, 333 and 999. (b) $37\,\overline{)\,6\,}$.

6. (a) $\sqrt{2}$; 8 slices. (b) $\dfrac{1}{\sqrt{2}}$; 4 slices.

7. (a) 680. (b) $\{715, 364, 924\}$ or $\{-286, 364, -77\}$.

8. (a) $\angle BPC = \frac{\pi}{2}$, $\angle CPA = \frac{5\pi}{6}$, $\angle APB = \frac{2\pi}{3}$. (b) $\overline{BC} = \sqrt{7}\ cm$.

1994–1995 First Round (November 12, 1994)

1. 6 weeks.

2. If the vertex C is on the major arc $\overset{\frown}{AB}$ (i.e., if the vertex C and the circumcenter O are on the same sides of the chord AB), then $\angle C = 27^0$; but if the vertex C is on the minor arc $\overset{\frown}{AB}$ (i.e., the vertex C and the circumcenter O are on the opposite sides of the chord AB), then $\angle C = 153^0$.

3. (a) $x^4 + \dfrac{1}{x^4} = y^4 - 4y^2 + 2$. (b) $x^5 + \dfrac{1}{x^5} = y^5 - 5y^3 + 5y$.

4. (a) 56. (b) 24.

5. (a) $h = \dfrac{1}{2}$, $k = -\dfrac{1}{2}$. (b) The sum is $\dfrac{36}{55}$.

6. $\overline{KD} = 6$ in both cases.

7. $a = 42,\ b = 7,\ c = 3,\ p = 5$.

8. $\overline{AB} = 2\sqrt{3}$.

1994–1995 Final Round (February 4, 1995)

1. The (minimum) number of 'foldings' needed is 29.

2. (a) 76. (b) 24.

3. (a) $y = 4$ or $y = -6$. (b) $x = 2 \pm \sqrt{3}$ or $x = -3 \pm 2\sqrt{2}$.

4. $$\frac{2ab}{a+b} < \sqrt{ab} < \frac{a+b}{2} < \sqrt{\frac{a^2+b^2}{2}}.$$

5. (a) $a_{19} = 190$. (b) $\displaystyle\sum_{n=1}^{95} \frac{1}{a_n} = \frac{95}{48}$

6. The diameter is 9.

7. (a) 5. (b) The difference obtained is always divisible by 9.

8. (a) $\angle APB = \frac{3\pi}{4}$. (b) $\overline{AB} = \sqrt{29}$.

1995–1996 First Round (November 11, 1995)

1. (a) $\{95, 1995\}$, $\{285, 665\}$.
   (b) $\{95, 285, 665\}$, $\{95, 285, 1995\}$, $\{95, 665, 1995\}$, $\{285, 665, 1995\}$.
   (c) $\{95, 285, 665, 1995\}$.

2. (a) 48. (b) The remaining eight triangles are equilateral with side length $\sqrt{2}$.

3. $(x, y) = (3612, 3613), (720, 725), (204, 221), (132, 157)$.

4. The area of the parallelogram is $5\sqrt{3}\ cm^2$.

5. $-\dfrac{14}{15}$.

6. $8\sqrt{2}\ cm$ for the external tangents, and $2\sqrt{11}\ cm$ for the internal tangents.

7. (a)

$$\begin{aligned} g(x) &= \frac{1+x^2}{(1-x+x^2)(1+x+x^2)} = \frac{1+x^2}{1+x^2+x^4}, \\ h(x) &= \frac{x}{(1-x+x^2)(1+x+x^2)} = \frac{x}{1+x^2+x^4}. \end{aligned}$$

(b) Yes, such a decomposition is unique.

8. (a) $a = 2$, $b = 1$. (b) $c = 2$, $d = 3$. (c) $p = 3$, $q = 2$, $r = 2$, $s = 1$.

1995–1996 Final Round (February 3, 1996)

1. $166\ cm$.

2. The other two sides are CE and EG, or CM and IG.

3. The minimum value is $\sqrt{\dfrac{24}{6}} + \sqrt{\dfrac{96}{24}} = 4$.

4. $n = 11$.

5. (a) $\alpha^2 + \beta^2 + \gamma^2 = 4$. (b) $\alpha^3 + \beta^3 + \gamma^3 = -9$.

6. $\overline{AD} = 10$, $\overline{AE} = 12$; or $\overline{AD} = 12$, $\overline{AE} = 10$.

7. (a) $g(x) = \dfrac{x}{\sqrt{(x^2+3)(1+3x^2)}}$, $h(x) = \sqrt{\dfrac{1+3x^2}{x^2+3}}$.

   (b) Yes, such a decomposition of a function f is unique (provided $f(x) > 0$ whenever $x > 0$).

8. (a) $a = 1$, $b = 2$.

   (d) $k_5 = 82$.

1997–1998 First Round (November 15, 1997)

1. $x = 1$, $y = 1$, $z = 3$.

2. The length of a side is $\sqrt{29}\ cm$.

3. $p = q = \dfrac{3}{5}$, $r = -\dfrac{1}{5}$.

4. $[ABCD] = 74\ cm^2$.

5. (a) $a = 2$, $b = -11$. (b) $c = 4$, $d = -1$.

6. (a) 72 ways. (b) 420 ways.

7. (a) $uv = -2$. (b) $u^3 + v^3 = 45$.

8. (a) $\overline{BC} = 8\ cm$. (b) $[ABC] = 10\sqrt{3}\ cm^2$.

1997–1998 Final Round (February 7, 1998)

1. The product is 1.

2. $[ABCD] = 49\ cm^2$.

3. $x = 3$, $y = 2$, $u = 8$, $v = 1$.

4. (a) $\sin\theta \cdot \cos\theta = -\frac{1}{8}$. (b) $\sin^3\theta - \cos^3\theta = \frac{7\sqrt{5}}{16}$.

5. $k = 97$.

6. (a) $[ABCD] = 12\ cm^2$. (b) $\overline{AB} = \sqrt{2}\ cm$, $\overline{BC} = 6\sqrt{2}\ cm$.

7. $p = 5$, $q = 4$, $r = 6$, $s = 5$.

8. (a) 30 ways. (b) 180 ways.

1998–1999 First Round (November 14, 1998)

1. The value of the expression is 1.

2. 7 points. One point at each of the six vertices of a regular hexagon inscribed in a unit circle, and a point at the center.

3. (a) $x^2 + y^2 = 98$. (b) $x^3 + y^3 = 970$.

4. The area is $32\ cm^2$.

5. (a) $s_1 s_2 - s_3$. (b) $s_1^3 - 3s_1 s_2 + 3s_3$.

6. (a) $R = \frac{5}{2}$. (b) $r = 1$.

7. (a) $f(x) = \frac{x}{x-1}$. (b) $g_n(x) = \left(\frac{x^2}{x-1}\right)^n$ (for any positive integer n) will do.

8. $\overline{BC} = 11\ cm$.

1998–1999 Final Round (February 6, 1999)

1. $p = 1997$.

2. (a) $(-16, 0)$, $(9, 0)$, $(0, 12)$. (b) $(0, 0)$, $(3, 3)$, $(-4, 4)$.
   Naturally, there exist infinitely many solutions to both (a) and (b).

3. 1089.

4. (a) $\varphi = \arctan \frac{4}{3}$ $(0 < \varphi < \frac{\pi}{2})$. (b) $|k| \geq \sqrt{3}$.

5. The value of the expression is 0.

6. $[ABC] = 294\, cm^2$.

7. The multiplicity of the root $x = 1$ is a positive even number.

8. (a) $[ABC] = r(2R + r)$. (b) $2R$, $(R + r) \pm \sqrt{R^2 - 2Rr - r^2}$,
   where R and r are the circumradius and the inradius.

APPENDIX C

Calculus Competitions

(Brief Solutions in Appendix D)

April 20, 1985

1. Find $\dfrac{d^{100}y}{dx^{100}}$, where $y = \dfrac{1}{1-x^2}$.

2. Suppose a polynomial p satisfies

$$x^2 p''(x) + (1-x)p'(x) - 3p(x) = 0, \quad p(0) = 1.$$

   (a) Find the degree of p.

   (b) Find p.

3. (a) Determine the constant a such that $f(x) = e^{ax}\sin x$ $(x > 0)$ has a local maximum at $x = \dfrac{\pi}{4}$.

   (b) With the constant a determined in Part (a), find the sum of all the local maximum values of f. Simplify your answer.

4. Is there a quadratic polynomial f satisfying the following two conditions?

   (i) $f(0) = 3$.

   (ii) $\displaystyle\int_0^3 f(x)g(x)\,dx = 0$ for every polynomial g of degree 1 or 0.

5. Let f be a continuous function in the interval $[0,\, 1]$.

   (a) Show that $\displaystyle\int_0^\pi x f(\sin x)\,dx = \frac{\pi}{2}\int_0^\pi f(\sin x)\,dx.$

   (b) Evaluate $\displaystyle\int_0^\pi \frac{x\sin x}{1+\cos^2 x}dx.$

6. Find the constants a, b and c such that

$$\lim_{x\to 1}\frac{ax^4 + bx^3 + 1}{(x-1)\sin \pi x} = c.$$

April 23, 1988

1. Evaluate the definite integrals:

   (a) $\int_3^7 (x-1)(x-2)^2(x-3)^3(x-4)^4(x-5)^5(x-6)^4(x-7)^3(x-8)^2(x-9)\,dx.$

   (b) $\int_0^{\frac{\pi}{2}} \frac{\sin x}{\sin x + \cos x} dx.$

2. Suppose $\lim_{x\to\infty} f(x) = \infty$ and $\lim_{x\to\infty} g(x) = \infty$. Prove or disprove:

   (a) If $\lim\limits_{x\to\infty} \frac{f(x)}{g(x)} = 0$, then $\lim\limits_{x\to\infty} \frac{e^{f(x)}}{e^{g(x)}} = 0$.

   (b) If $\lim\limits_{x\to\infty} \frac{f(x)}{g(x)} = 1$, then $\lim\limits_{x\to\infty} \frac{e^{f(x)}}{e^{g(x)}} = 1$.

   (c) If $\lim\limits_{x\to\infty} \frac{f(x)}{g(x)} = 0$, then $\lim\limits_{x\to\infty} \frac{\log f(x)}{\log g(x)} = 0$.

   (d) If $\lim\limits_{x\to\infty} \frac{f(x)}{g(x)} = 1$, then $\lim\limits_{x\to\infty} \frac{\log f(x)}{\log g(x)} = 1$.

   Here, log denotes the natural logarithm.

3. Show that
$$\frac{2}{5} < \int_0^1 x^{\sin x + \cos x} dx < \frac{1}{2}.$$

4. For a cubic polynomial f with real coefficients, if $f(x) = 0$ has three distinct positive roots, prove that
$$f(x) + xf'(x) = 0$$
also has three distinct positive roots.

5. Find the most general form of functions f satisfying
$$\int_a^b f(t)\,dt = (b-a)\cdot f\left(\frac{a+b}{2}\right) \quad \text{for all real numbers } a \text{ and } b.$$
You may impose additional conditions on the function f if you deem it necessary, but be sure to indicate clearly what conditions you have imposed.

6. Find the integral part n of
$$T = \frac{1}{\sqrt{1}} + \frac{1}{\sqrt{2}} + \cdots + \frac{1}{\sqrt{99}};$$
i.e., find the integer n such that $n \le T < n+1$.

7. Find the function f satisfying
$$f(x) = \sin x + \int_0^{\pi} f'(t)\cos(x-t)\,dt.$$

October 8, 1988

1. Find the area of a triangle whose orthogonal projections to the three coordinate planes have areas 12, 16 and 21, respectively.

2. Suppose a sequence $\{a_n\}_{n=1}^{\infty}$ has the property that

$$a_1 + a_2 + \cdots + a_n = \frac{1}{6}n(n+1)(n+2) \quad \text{for all } \; n = 1, 2, 3, \cdots .$$

   Find the sum of the infinite series $\sum_{n=1}^{\infty} \frac{1}{a_n}$.

3. Evaluate the improper integral

$$\int_0^{\infty} \frac{\log x}{1 + x + x^2} dx,$$

   where log denotes the natural logarithm.

4. Find the sum of of the infinite series

$$\frac{1^2}{3} - \frac{2^2}{3^2} + \frac{3^2}{3^3} - \frac{4^2}{3^4} + \cdots + (-1)^{k-1}\frac{k^2}{3^k} + \cdots .$$

5. For each positive integer n, let S_n be the sum of all the positive integer divisors of 1988^n (including 1 and 1988^n itself). For example, $S_1 = 4032$. Find the limit

$$\lim_{n \to \infty} \frac{S_n}{1988^n}.$$

6. Show that, for an arbitrary positive integer n,

$$\sum_{k=0}^{n} \frac{(-1)^k}{2k+1} \binom{n}{k} = \frac{2 \cdot 4 \cdots (2n-2) \cdot (2n)}{3 \cdot 5 \cdots (2n-1) \cdot (2n+1)}.$$

   Hint: Why is this a problem in calculus?

April 8, 1989

1. (a) Show that the cubic equation $x^3 - x^2 - 3 = 0$ has a unique (real) root between 1 and 2.

   (b) Show that this root is an irrational number.

2. Find the polynomial p of least degree simultaneously satisfying

$$\lim_{x\to 0} \frac{p(x)}{x} = 2, \quad \lim_{x\to -1} \frac{p(x)}{x+1} = 3, \quad \lim_{x\to 2} \frac{p(x)}{x-2} = 6.$$

3. (a) Show that there exists an angle θ such that

$$3\cos x + 4\sin x = 5\sin(x+\theta) \quad \text{for all real number } x.$$

   (b) Evaluate the integral

$$\int_0^\pi |a\cos x + b\sin x|\, dx,$$

   where a and b are constants.

4. Let

$$f(x) = \begin{cases} 1 - |x| & (\text{for} \quad |x| \le 1); \\ 0 & (\text{for} \quad |x| > 1). \end{cases}$$

   (a) Sketch the graph of $y = 3f(3x)$.

   (b) Evaluate $\lim_{n\to\infty} \displaystyle\int_{-1}^{1} nf(nx)\cos x\, dx$.

5. It is well known that

$$\lim_{n\to\infty} \left(1 + \frac{1}{2} + \frac{1}{3} + \cdots + \frac{1}{n} - \log n\right) = c,$$

where log denotes the natural logarithm and $c = 0.5772156649\cdots$ is the Euler constant. Use this result to find the constant k for which the limit

$$\lim_{n\to\infty} \left(1 + \frac{1}{3} + \frac{1}{5} + \cdots + \frac{1}{2n-1} - k\log n\right)$$

exists, and find this limit in terms of c.

6. (a) Let $\{c_n\}_{n=1}^{\infty}$ be a sequence of positive numbers such that

$$(c_1 c_2 \cdots c_n)^{1/n} = n + 1 \quad \text{for all natural number } n.$$

   Evaluate

$$\lim_{n\to\infty} \frac{c_n}{n}.$$

(b) We want to prove the following theorem of T. Carlemann: If the series with positive terms

$$a_1 + a_2 + a_3 + \cdots + a_n + \cdots$$

is convergent, then the series

$$a_1 + (a_1 a_2)^{1/2} + (a_1 a_2 a_3)^{1/3} + \cdots + (a_1 a_2 a_3 \cdots a_n)^{1/n} + \cdots$$

is also convergent.

Here is a proof due to G. Pólya. With $\{c_n\}_{n=1}^{\infty}$ to be determined later, we proceed as follows: By the inequality between the arithmetic mean and the geometric mean, we have

$$\begin{aligned}\sum_{n=1}^{\infty} (a_1 a_2 a_3 \cdots a_n)^{1/n} &= \sum_{n=1}^{\infty} \frac{(a_1 c_1 \cdot a_2 c_2 \cdots a_n c_n)^{1/n}}{(c_1 c_2 \cdots c_n)^{1/n}} \\ &\leq \sum_{n=1}^{\infty} \frac{a_1 c_1 + a_2 c_2 + \cdots + a_n c_n}{n\,(c_1 c_2 \cdots c_n)^{1/n}} \\ &= \sum_{k=1}^{\infty} \left\{ a_k c_k \sum_{n=k}^{\infty} \frac{1}{n\,(c_1 c_2 \cdots c_n)^{1/n}} \right\}.\end{aligned}$$

Complete the proof with a suitable choice of $\{c_n\}_{n=1}^{\infty}$.

7. (a) Compute the first few cases of

$$I_n = \int_0^{\pi} \left(\frac{\sin nx}{\sin x} \right) dx \quad (n = 1,\, 2,\, \cdots);$$

then guess and prove the general case.

(b) Evaluate the integrals

$$J_n = \int_0^{\pi} \left(\frac{\sin nx}{\sin x} \right)^2 dx \quad (n = 1,\, 2,\, \cdots).$$

APPENDIX D

Brief Solutions to Calculus Competitions

April 20, 1985

1.

$$\begin{aligned} y &= \frac{1}{1-x^2} = \frac{1}{2}\left(\frac{1}{1-x} + \frac{1}{1+x}\right) \\ \frac{d^n y}{dx^n} &= \frac{n!}{2}\left\{\frac{1}{(1-x)^{n+1}} + \frac{(-1)^n}{(1+x)^{n+1}}\right\}. \end{aligned}$$

2. (a) Let $n = \deg p$. Then the leading coefficient in the given equation gives

$$n(n-1) - n - 3 = n^2 - 2n - 3 = (n-3)(n+1) = 0.$$

Because $n \geq 0$, we obtain $n = 3$.

(b) Let $p(x) = ax^3 + bx^2 + cx + 1$. Then

$$\begin{aligned} &x^2(6ax + 2b) + (1-x)(3ax^2 + 2bx + c) \\ &\quad - 3(ax^3 + bx^2 + cx + 1) = 0. \end{aligned}$$

Comparing the coefficients of x^3, x^2, x^1, x^0, we obtain

$$\begin{cases} 6a - 3a - 3a = 0, \\ 2b + 3a - 2b - 3b = 0, \\ 2b - c - 3c = 0, \\ c - 3 = 0. \end{cases}$$

Solving this system of simultaneous equations, we obtain $a = b = 6$, $c = 3$.

$$\therefore\ p(x) = 6x^3 + 6x^2 + 3x + 1.$$

3. (a)

$$\begin{aligned} f(x) &= e^{ax} \cdot \sin x, \\ f'(x) &= e^{ax}\{a \sin x + \cos x\}. \\ \therefore\ f'\left(\frac{\pi}{4}\right) &= e^{\frac{\pi a}{4}}\{a+1\}\frac{\sqrt{2}}{2} = 0 \end{aligned}$$

if and only if $a = -1$. It is simple to verify that $f(x) = e^{-x} \cdot \sin x$ has a local maximum at $x = \frac{\pi}{4}$. Furthermore, $f(x) = e^{-x} \cdot \sin x$ $(x > 0)$ has a local maximum if and only if $x = 2n\pi + \frac{\pi}{4}$.

$$\begin{aligned}\sum_{n=0}^{\infty} f\left(2n\pi + \frac{\pi}{4}\right) &= \frac{\sqrt{2}}{2} \cdot \sum_{n=0}^{\infty} e^{-(2n\pi + \frac{\pi}{4})} \\ &= \frac{\sqrt{2}}{2} \cdot \frac{e^{-\frac{\pi}{4}}}{1 - e^{-2\pi}}.\end{aligned}$$

4. Let $f(x) = ax^2 + bx + 3$. Then we must have

$$\begin{aligned}\int_0^3 (ax^2 + bx + 3) \cdot 1\, dx &= \left[\frac{ax^3}{3} + \frac{bx^2}{2} + 3x\right]_0^3 \\ &= 9\left(a + \frac{b}{2} + 1\right) = 0; \\ \int_0^3 (ax^2 + bx + 3) \cdot x\, dx &= \left[\frac{ax^4}{4} + \frac{bx^3}{3} + \frac{3x^2}{2}\right]_0^3 \\ &= 9\left(\frac{9a}{4} + b + \frac{3}{2}\right) = 0.\end{aligned}$$

$$\therefore\ a = 2, \quad b = -6; \qquad f(x) = 2x^2 - 6x + 3.$$

It is simple to verify that this quadratic polynomial satisfies the condition.

Exercise. (a) Find a cubic polynomial f such that

$$\int_{-1}^{1} f(x)g(x)\, dx = 0$$

for every polynomial g of degree at most 2.

(b) Generalize.

5. (a) Let $x = \pi - t$. Then

$$\begin{aligned}\int_0^{\pi} x \cdot f(\sin x) &= -\int_{\pi}^{0} (\pi - t) \cdot f(\sin(\pi - t))\, dt \\ &= \pi \int_0^{\pi} f(\sin t)\, dt - \int_0^{\pi} t \cdot f(\sin t)\, dt.\end{aligned}$$

Because the second term on the right is identical to the left-most member, we obtain

$$\int_0^{\pi} x \cdot f(\sin x)\, dx = \frac{\pi}{2} \int_0^{\pi} f(\sin x)\, dx.$$

(b)

$$\int_0^{\pi} \frac{x \sin x}{1 + \cos^2 x} dx = \frac{\pi}{2} \int_0^{\pi} \frac{\sin x}{1 + \cos^2 x} dx \quad (u = \cos x)$$

$$= \frac{\pi}{2}\int_1^{-1}\frac{-du}{1+u^2} = \frac{\pi}{2}[\arctan u]_{-1}^{1}$$
$$= \frac{\pi}{2}\left(\frac{\pi}{4}+\frac{\pi}{4}\right) = \frac{\pi^2}{4}.$$

6. Note that $f(x) = ax^4 + bx^3 + 1$ must have a double root at $x = 1$.

$$f(1) = a+b+1 = 0, \quad f'(1) = 4a+3b = 0. \qquad \therefore\ a = 3,\ b = -4.$$

$$\begin{aligned}
\lim_{x\to 1}\frac{ax^4+bx^3+1}{(x-1)\sin\pi x} &= \lim_{x\to 1}\frac{3x^4-4x^3+1}{(x-1)\sin\pi x} \\
&= \lim_{x\to 1}\frac{(x-1)(3x^3-x^2-x-1)}{(x-1)\sin\pi x} \\
&= \lim_{x\to 1}\frac{3x^3-x^2-x-1}{\sin\pi x} \\
&= \lim_{x\to 1}\frac{9x^2-2x-1}{\pi\cos\pi x} = -\frac{6}{\pi}.
\end{aligned}$$

April 23, 1988

1. (a) Let $x = t + 5$. Then

$$\begin{aligned}
&\int_3^7 (x-1)(x-2)^2(x-3)^3(x-4)^4(x-5)^5(x-6)^4(x-7)^3(x-8)^2(x-9)\,dx \\
&= \int_{-2}^{2}(t+4)(t+3)^2(t+2)^3(t+1)^4t^5(t-1)^4(t-2)^3(t-3)^2(t-4)\,dt \\
&= \int_{-2}^{2}(t^2-4^2)(t^2-3^2)^2(t^2-2^2)^3(t^2-1^2)^4t^5\,dt \\
&= 0 \quad (\because \text{ the integrand is an odd function.})
\end{aligned}$$

Alternatively, let $x = 10 - t$, then

$$\begin{aligned}
I &= \int_3^7 (x-1)(x-2)^2(x-3)^3(x-4)^4(x-5)^5(x-6)^4(x-7)^3(x-8)^2(x-9)\,dx \\
&= -\int_7^3 (9-t)(8-t)^2(7-t)^3(6-t)^4(5-t)^5(4-t)^4(3-t)^3(2-t)^2(1-t)\,dt \\
&= -\int_3^7 (t-9)(t-8)^2(t-7)^3(t-6)^4(t-5)^5(t-4)^4(t-3)^3(t-2)^2(t-1)\,dt \\
&= -I. \quad \therefore\ I = 0.
\end{aligned}$$

(b) Let $x = \frac{\pi}{2} - t$. Then

$$I = \int_0^{\frac{\pi}{2}}\frac{\sin x}{\sin x+\cos x}dx$$

$$
\begin{aligned}
&= -\int_{\frac{\pi}{2}}^{0} \frac{\sin(\frac{\pi}{2}-t)}{\sin(\frac{\pi}{2}-t)+\cos(\frac{\pi}{2}-t)}dt \\
&= \int_0^{\frac{\pi}{2}} \frac{\cos t}{\cos t+\sin t}dt. \\
\therefore\ 2I &= \int_0^{\frac{\pi}{2}} \frac{\sin x+\cos x}{\sin x+\cos x}dx = \int_0^{\frac{\pi}{2}} dx = \frac{\pi}{2}. \qquad I = \frac{\pi}{4}.
\end{aligned}
$$

Exercise.. (a) Show that

$$\int_0^{\frac{\pi}{2}} \sin^2 x\,dx = \int_0^{\frac{\pi}{2}} \cos^2 x\,dx = \frac{\pi}{4}.$$

(b) Evaluate

$$\int_0^{\frac{\pi}{2}} \frac{\sin^3 x}{\sin x+\cos x}dx.$$

Remark. Of course, both (a) and (b) in the exercise can be evaluated by a usual method, but it is instructive to use the idea employed in 1(b) above.

2. (a) True. (b) False. (c) False. (d) True.

(a) Because $\lim_{x\to\infty} \dfrac{f(x)}{g(x)} = 0$, we have $f(x) < \frac{1}{2}g(x)$ for all x sufficiently large.

$$\therefore\ 0 \le \lim_{x\to\infty} \frac{e^{f(x)}}{e^{g(x)}} = \lim_{x\to\infty} e^{f(x)-g(x)} \le \lim_{x\to\infty} e^{(\frac{1}{2}-1)g(x)} = 0.$$

(b) Let $f(x) = x+1$, $g(x) = x$. Then

$$\lim_{x\to\infty} \frac{e^{f(x)}}{e^{g(x)}} = \lim_{x\to\infty} e^{(x+1)-x} = e.$$

(c) Let $f(x) = x$, $g(x) = x^2$. Then

$$\lim_{x\to\infty} \frac{\log f(x)}{\log g(x)} = \lim_{x\to\infty} \frac{\log x}{2\log x} = \frac{1}{2}.$$

(d)

$$
\begin{aligned}
\lim_{x\to\infty} \frac{\log f(x)}{\log g(x)} - 1 &= \lim_{x\to\infty} \frac{\log f(x) - \log g(x)}{\log g(x)} \\
&= \lim_{x\to\infty} \frac{\log\left\{\frac{f(x)}{g(x)}\right\}}{\log g(x)} = 0,
\end{aligned}
$$

because $\lim_{x\to\infty} \log \dfrac{f(x)}{g(x)} = \log 1 = 0$, and $\lim_{x\to\infty} \log g(x) = \infty$.

3. Note that $\sin x + \cos x = \sqrt{2} \cdot \sin\left(x + \frac{\pi}{4}\right)$. Hence, for $0 \le x \le 1$, we have $\frac{\pi}{4} \le x + \frac{\pi}{4} < \frac{\pi}{2} + \frac{\pi}{4}$.

$$\begin{aligned}
\therefore\ 1 = \sqrt{2} \cdot \sin \frac{\pi}{4} &\le \sin x + \cos x \le \sqrt{2} < \frac{3}{2}. \\
x^{3/2} &\le x^{\sin x + \cos x} \le x \quad (0 \le x \le 1). \\
\int_0^1 x^{3/2} dx &< \int_0^1 x^{\sin x + \cos x} dx < \int_0^1 x\, dx; \\
\therefore\ \frac{2}{5} &< \int_0^1 x^{\sin x + \cos x} dx < \frac{1}{2}.
\end{aligned}$$

4. Without loss of generality, we may assume the leading coefficient is 1. Set

$$\begin{aligned}
f(x) &= (x - \alpha)(x - \beta)(x - \gamma) \quad (0 < \alpha < \beta < \gamma), \\
g(x) &= f(x) + x f'(x).
\end{aligned}$$

Then

$$\begin{aligned}
g(0) &= -\alpha\beta\gamma < 0, \\
g(\alpha) &= \alpha(\alpha - \beta)(\alpha - \gamma) > 0, \\
g(\beta) &= \beta(\beta - \alpha)(\beta - \gamma) < 0, \\
g(\gamma) &= \gamma(\gamma - \alpha)(\gamma - \beta) > 0.
\end{aligned}$$

Therefore, $g(x) = 0$ has at least one root in each of the intervals $(0, \alpha)$, (α, β), (β, γ), but because g is of degree 3, there cannot be more than one root in each interval.

Exercise. (a) For a cubic polynomial f with real coefficients, if f has three distinct real roots, and $f(0) \ne 0$, prove that

$$f(x) + x f'(x) = 0$$

also has three distinct real roots.

(b) Can this result be extended to polynomials of higher degree?

5. Assuming that the function f is differentiable, let us differentiate both sides of the given equality

$$\int_a^b f(t)\, dt = (b - a) \cdot f\left(\frac{a+b}{2}\right)$$

with respect to b; we obtain

$$f(b) = f\left(\frac{a+b}{2}\right) + \frac{b-a}{2} \cdot f'\left(\frac{a+b}{2}\right).$$

Similarly, differentiating with respect to a (or interchanging the roles of a and b in the last equality), we obtain

$$f(a) = f\left(\frac{a+b}{2}\right) - \frac{b-a}{2} \cdot f'\left(\frac{a+b}{2}\right).$$

Adding the last two equalities, we obtain

$$\frac{f(a)+f(b)}{2} = f\left(\frac{a+b}{2}\right).$$

Differentiating both sides with respect to a, we obtain

$$f'(a) = f'\left(\frac{a+b}{2}\right).$$

Substituting $a = 0$, this equality becomes

$$f'(0) = f'\left(\frac{b}{2}\right).$$

Because this relation has to be true for all b, we conclude that f' must be a constant, and so f is a polynomial of degree at most 1.

Conversely, it is simple to verify that any polynomial of degree at most 1 satisfies the given equation.

6. Note that $f(x) = \dfrac{1}{\sqrt{x}}$ is a monotone decreasing function for $x > 0$.

$$\begin{aligned}
\therefore \int_k^{k+1} \frac{dx}{\sqrt{x}} &< \frac{1}{\sqrt{k}} < \int_{k-1}^{k} \frac{dx}{\sqrt{x}}; \\
2\left(\sqrt{k+1}-\sqrt{k}\right) &< \frac{1}{\sqrt{k}} < 2\left(\sqrt{k}-\sqrt{k-1}\right) \quad (k = 1,\ 2,\ \cdots). \\
\sum_{k=1}^{99} \frac{1}{\sqrt{k}} &= \frac{1}{\sqrt{1}} + \sum_{k=2}^{99} \frac{1}{\sqrt{k}} \\
&< 1 + 2\sum_{k=2}^{99}\left(\sqrt{k}-\sqrt{k-1}\right) \\
&= 1 + 2\left(\sqrt{99}-\sqrt{1}\right) \\
&< 1 + 2\left(\sqrt{100}-1\right) = 19. \\
\sum_{k=1}^{99} \frac{1}{\sqrt{k}} &> 2\sum_{k=1}^{99}\left(\sqrt{k+1}-\sqrt{k}\right) \\
&= 2\left(\sqrt{100}-1\right) = 18. \\
\therefore 18 &< \sum_{k=1}^{99} \frac{1}{\sqrt{k}} < 19.
\end{aligned}$$

7.

$$f(x) = \sin x + \int_0^{\pi} f'(t)\cos(x-t)\,dt$$

$$\begin{aligned} &= \sin x + \int_0^{\pi} f'(t)(\cos x \cdot \cos t + \sin x \cdot \sin t)\,dt \\ &= \cos x \int_0^{\pi} f'(t) \cos t\,dt + \sin x \left(1 + \int_0^{\pi} f'(t) \sin t\,dt\right) \\ &= a \cos x + b \sin x, \end{aligned}$$

where $a = \int_0^{\pi} f'(t) \cos t\,dt,\ b = 1 + \int_0^{\pi} f'(t) \sin t\,dt.$

$$\begin{aligned} a &= \int_0^{\pi} f'(t) \cos t\,dt \\ &= \int_0^{\pi} (-a \sin t + b \cos t) \cos t\,dt \\ &= b \int_0^{\pi} \cos^2 t\,dt = \frac{b\pi}{2}. \\ b &= 1 + \int_0^{\pi} (-a \sin t + b \cos t) \sin t\,dt \\ &= 1 - a \int_0^{\pi} \sin^2 t\,dt = 1 - \frac{a\pi}{2}. \end{aligned}$$

$$\therefore\ a - \frac{b\pi}{2} = 0, \quad \frac{a\pi}{2} + b = 1; \qquad a = \frac{2\pi}{\pi^2 + 4}, \quad b = \frac{4}{\pi^2 + 4}.$$

$$\therefore\ f(x) = \frac{2}{\pi^2 + 4}(\pi \cos x + 2 \sin x).$$

October 8, 1988

1. Let $\cos\alpha$, $\cos\beta$, $\cos\gamma$ be the direction cosines of the perpendicular to the plane of the triangle, and S the area of the triangle. Then

$$S \cos\alpha = 12, \quad S \cos\beta = 16, \quad S \cos\gamma = 21.$$

Because $\cos^2\alpha + \cos^2\beta + \cos^2\gamma = 1$, we have

$$S = \sqrt{12^2 + 16^2 + 21^2} = 29.$$

2.

$$\begin{aligned} a_n &= (a_1 + a_2 + \cdots + a_n) - (a_1 + a_2 + \cdots + a_{n-1}) \\ &= \frac{1}{6}n(n+1)(n+2) - \frac{1}{6}(n-1)n(n+1) \\ &= \frac{1}{6}n(n+1)\{(n+2) - (n-1)\} = \frac{1}{2}n(n+1). \\ \therefore\ \sum_{n=1}^{\infty} \frac{1}{a_n} &= \sum_{n=1}^{\infty} \frac{2}{n(n+1)} = 2\sum_{n=1}^{\infty}\left(\frac{1}{n} - \frac{1}{n+1}\right) = 2. \end{aligned}$$

October 8, 1988

3.

$$\begin{aligned}\int_0^\infty \frac{\log x}{1+x+x^2}dx &= \int_0^1 \frac{\log x}{1+x+x^2}dx + \int_1^\infty \frac{\log x}{1+x+x^2}dx \\ &= \int_0^1 \frac{\log x}{1+x+x^2}dx - \int_1^0 \frac{\log(\frac{1}{t})}{1+(\frac{1}{t})+(\frac{1}{t})^2}\cdot\frac{dt}{t^2} \quad \left(x=\frac{1}{t}\right) \\ &= \int_0^1 \frac{\log x}{1+x+x^2}dx - \int_0^1 \frac{\log t}{1+t+t^2}dt = 0.\end{aligned}$$

Remark. The reader should verify the convergence of this improper integral.

Exercise. Evaluate

$$\int_0^\infty \frac{\log x}{a^2+ax+x^2}dx \quad (a>0).$$

4. We start from the geometric series

$$1+x+x^2+\cdots+x^k+\cdots = \frac{1}{1-x} \quad (|x|<1).$$

Differentiate both sides with respect to x and then multiply by x; we obtain

$$1x+2x^2+3x^3+\cdots+kx^k+\cdots = \frac{x}{(1-x)^2} \quad (|x|<1).$$

Repeating this procedure once more, we obtain

$$\begin{aligned}&1^2x+2^2x^2+3^2x^3+\cdots+k^2x^k+\cdots \\ &= x\left\{\frac{1}{(1-x)^2}+\frac{2x}{(1-x)^3}\right\} \\ &= \frac{x(1+x)}{(1-x)^3} \quad (|x|<1).\end{aligned}$$

Substituting $x=-\frac{1}{3}$ and changing the sign of both sides, we obtain

$$\frac{1^2}{3}-\frac{2^2}{3^2}+\frac{3^2}{3^3}-\frac{4^2}{3^4}+\cdots+(-1)^{k-1}\frac{k^2}{3^k}+\cdots = \frac{\frac{1}{3}\cdot\frac{2}{3}}{\left(\frac{4}{3}\right)^3} = \frac{3}{32}.$$

5.

$$\begin{aligned}1988^n &= 2^{2n}\cdot 7^n\cdot 71^n. \\ S_n &= \left(1+2+2^2+\cdots+2^{2n}\right)\cdot\left(1+7+7^2+\cdots+7^n\right) \\ &\quad \cdot\left(1+71+71^2+\cdots+71^n\right) \\ &= \frac{2^{2n+1}-1}{2-1}\cdot\frac{7^{n+1}-1}{7-1}\cdot\frac{71^{n+1}-1}{71-1} \\ \lim_{n\to\infty}\frac{S_n}{1988^n} &= \frac{1}{6\cdot 70}\lim_{n\to\infty}\left(\frac{2^{2n+1}-1}{2^{2n}}\cdot\frac{7^{n+1}-1}{7^n}\cdot\frac{71^{n+1}-1}{71^n}\right) \\ &= \frac{2\cdot 7\cdot 71}{6\cdot 70} = \frac{71}{30}.\end{aligned}$$

6.
$$\begin{aligned}\sum_{k=0}^{n}\frac{(-1)^k}{2k+1}\binom{n}{k} &= \sum_{k=0}^{n}(-1)^k\binom{n}{k}\int_0^1 x^{2k}dx\\ &= \int_0^1\left\{\sum_{k=0}^{n}(-1)^k\binom{n}{k}x^{2k}\right\}dx\\ &= \int_0^1 (1-x^2)^n\,dx\\ &= \int_0^{\pi/2}\cos^{2n+1}\theta\,d\theta \qquad (x=\sin\theta)\\ &= \frac{2\cdot 4\cdot 6\cdots(2n)}{3\cdot 5\cdot 7\cdots(2n+1)},\end{aligned}$$

by the Wallis formula.

Exercise. Show that

$$\lim_{n\to\infty}\sqrt{n}\cdot\sum_{k=0}^{n}\frac{(-1)^k}{2k+1}\binom{n}{k}=\frac{\sqrt{\pi}}{2}.$$

April 8, 1989

1. (a) Let $f(x)=x^3-x^2-3$. Because $f(1)=-3<0$ and $f(2)=8-4-3>0$, there exists at least one root between 1 and 2, by the intermediate value theorem. Now $f'(x)=3x^2-2x=x(3x-2)>0$ for $1\le x\le 2$; hence f is monotone increasing in the interval [1, 2], and so there cannot exist more than one root in this interval.

(b) Suppose $x=\frac{p}{q}$ (where p and q are relatively prime integers) is a root. Then

$$\left(\frac{p}{q}\right)^3-\left(\frac{p}{q}\right)^2-3=0. \quad \therefore\ \frac{p^3}{q}=p^2+3q^2.$$

Now the right-hand side is an integer, so q must divide p^3; but p and q are relatively prime, so we must have $q=1$. However, if $q=1$, then the equality above can be rewritten as

$$p^2(p-1)=3.$$

In particular, an integer root p must be a factor of the constant term 3; i.e., $p=\pm 1$, ± 3. It is obvious that none of these integer is a root, so the root must be irrational.

2. Clearly, p must have 0, -1, 2 as its (simple) roots. Let

$$p(x)=x(x+1)(x-2)(ax^2+bx+c).$$

Then

$$\begin{aligned}
\lim_{x\to 0}\frac{p(x)}{x} &= \lim_{x\to 0}(x+1)(x-2)(ax^2+bx+c)\\
&= -2c = 2. \quad \therefore\ c=-1.\\
\lim_{x\to -1}\frac{p(x)}{x+1} &= \lim_{x\to -1}x(x-2)(ax^2+bx+c)\\
&= 3(a-b+c)\\
&= 3(a-b-1)=3. \quad \therefore\ a-b=2.\\
\lim_{x\to 2}\frac{p(x)}{x-2} &= \lim_{x\to 2}x(x+1)(ax^2+bx+c)\\
&= 2\cdot 3\cdot(4a+2b+c)\\
&= 6(4a+2b-1)=6. \quad \therefore\ 2a+b=1.
\end{aligned}$$

It follows that

$$a=1, \quad b=-1; \qquad p(x)=x(x+1)(x-2)(x^2-x-1).$$

3. (a) Because $(\frac{3}{5})^2+(\frac{4}{5})^2=1$, there exists an angle θ such that

$$\begin{aligned}
\sin\theta &= \frac{3}{5}, \quad \cos\theta=\frac{4}{5}.\\
3\cos x+4\sin x &= 5\left(\frac{3}{5}\cos x+\frac{4}{5}\sin x\right)\\
&= 5(\sin\theta\cdot\cos x+\cos\theta\cdot\sin x)\\
&= 5\sin(\theta+x).
\end{aligned}$$

(b) Choose φ such that

$$\sin\varphi=\frac{a}{\sqrt{a^2+b^2}}, \quad \cos\varphi=\frac{b}{\sqrt{a^2+b^2}}.$$

Then

$$\begin{aligned}
\int_0^\pi |a\cos x+b\sin x|\,dx &= \sqrt{a^2+b^2}\int_0^\pi |\sin(x+\varphi)|\,dx\\
&= \sqrt{a^2+b^2}\int_\varphi^{\pi+\varphi}|\sin t|\,dt \quad (t=x+\varphi)\\
&= \sqrt{a^2+b^2}\int_0^\pi |\sin t|\,dt \quad (\because\ |\sin t| \text{ is } \pi\text{-periodic})\\
&= \sqrt{a^2+b^2}\int_0^\pi \sin t\,dt\\
&= \sqrt{a^2+b^2}\,[-\cos t]_0^\pi = 2\sqrt{a^2+b^2}.
\end{aligned}$$

4. (a)

$$y = 3f(3x) = \begin{cases} 3(1-|3x|) & (|3x| \leq 1); \\ 0 & (|3x| > 1). \end{cases}$$

(b)

$$\begin{aligned}
&\int_{-1}^{1} nf(nx)\cos x\,dx \\
&= n\int_{-\frac{1}{n}}^{\frac{1}{n}} (1-|nx|)\cos x\,dx \\
&= 2n\int_{0}^{\frac{1}{n}} (1-nx)\cos x\,dx \\
&= 2n\left\{[(1-nx)\sin x]_0^{\frac{1}{n}} + n\int_0^{\frac{1}{n}} \sin x\,dx\right\} \\
&= 2n^2[-\cos x]_0^{\frac{1}{n}} = 2n^2\left(1-\cos\frac{1}{n}\right) \\
&= 2\cdot\frac{1-\cos\frac{1}{n}}{\left(\frac{1}{n}\right)^2} \longrightarrow 1 \quad (\text{as } n\to\infty).
\end{aligned}$$

5.

$$\begin{aligned}
&\lim_{n\to\infty}\left(1+\frac{1}{3}+\frac{1}{5}+\cdots+\frac{1}{2n-1}-k\log n\right) \\
&= \lim_{n\to\infty}\left\{\left(1+\frac{1}{2}+\frac{1}{3}+\cdots+\frac{1}{2n}-\log(2n)\right)\right. \\
&\quad -\frac{1}{2}\left(1+\frac{1}{2}+\frac{1}{3}+\cdots+\frac{1}{n}-\log n\right) \\
&\quad \left.+\log(2n)-\frac{1}{2}\log n - k\log n\right\} \\
&= c-\frac{c}{2}+\lim_{n\to\infty}\left\{\log 2+\left(\frac{1}{2}-k\right)\log n\right\}.
\end{aligned}$$

So the limit exists if and only if $k=\frac{1}{2}$, and in this case the limit is $\dfrac{c}{2}+\log 2$.

6. (a)

$$\begin{aligned}
(c_1c_2\cdots c_n)^{1/n} &= n+1. \\
c_n &= \frac{(c_1c_2\cdots c_n)}{(c_1c_2\cdots c_{n-1})} = \frac{(n+1)^n}{n^{n-1}}. \\
\therefore\ \lim_{n\to\infty}\frac{c_n}{n} &= \lim_{n\to\infty}\frac{(n+1)^n}{n^n} = \lim_{n\to\infty}\left(1+\frac{1}{n}\right)^n = e.
\end{aligned}$$

(b) Choose $\{c_n\}_{n=1}^{\infty}$ such that $(c_1c_2\cdots c_n)^{1/n} = n+1$. Then

$$\begin{aligned}
\sum_{n=1}^{\infty}(a_1a_2\cdots a_n)^{1/n} &\leq \sum_{k=1}^{\infty}\left\{a_kc_k\sum_{n=k}^{\infty}\frac{1}{n(c_1c_2\cdots c_n)^{1/n}}\right\} \\
&= \sum_{k=1}^{\infty}\left\{a_kc_k\sum_{n=k}^{\infty}\frac{1}{n(n+1)}\right\} \\
&= \sum_{k=1}^{\infty}\frac{a_kc_k}{k} = \sum_{k=1}^{\infty}a_k\left(1+\frac{1}{k}\right)^k \\
&< e\sum_{k=1}^{\infty}a_k.
\end{aligned}$$

7. (a)

$$\begin{aligned}
I_1 &= \int_0^{\pi}\left(\frac{\sin x}{\sin x}\right)dx = \pi, \\
I_2 &= \int_0^{\pi}\left(\frac{\sin 2x}{\sin x}\right)dx \\
&= \int_0^{\pi}2\cos x\,dx = 2[\sin x]_0^{\pi} = 0, \\
I_3 &= \int_0^{\pi}\left(\frac{\sin 3x}{\sin x}\right)dx \\
&= \int_0^{\pi}(3-4\sin^2 x)\,dx = 3\pi - 4\cdot\frac{1}{2}\cdot\pi = \pi, \\
I_4 &= \int_0^{\pi}\left(\frac{\sin 4x}{\sin x}\right)dx \\
&= 2\int_0^{\pi}\frac{\sin 2x\cdot\cos 2x}{\sin x}dx = 4\int_0^{\pi}\cos x\cdot\cos 2x\,dx \\
&= 4\left\{\int_0^{\pi/2}\cos x\cdot\cos 2x\,dx + \int_{\pi/2}^{\pi}\cos x\cdot\cos 2x\,dx\right\} \quad (x=\pi-t) \\
&= 4\left\{\int_0^{\pi/2}\cos x\cdot\cos 2x\,dx + \int_{\pi/2}^{0}\cos t\cdot\cos 2t\,dt\right\} = 0.
\end{aligned}$$

Thus we conjecture that $I_n = \pi$ if n is odd, and $I_n = 0$ if n is even.

$$\begin{aligned}
I_{n+2} - I_n &= \int_0^{\pi}\frac{\sin(n+2)x - \sin nx}{\sin x}dx \\
&= 2\int_0^{\pi}\frac{\cos(n+1)x\cdot\sin x}{\sin x}dx \\
&= 2\int_0^{\pi}\cos(n+1)x\,dx = \frac{2}{n+1}[\sin(n+1)x]_0^{\pi} = 0.
\end{aligned}$$

(b)

$$\begin{aligned}
J_1 &= \int_0^\pi \left(\frac{\sin x}{\sin x}\right)^2 dx = \pi, \\
J_2 &= \int_0^\pi \left(\frac{\sin 2x}{\sin x}\right)^2 dx = 4\int_0^\pi \cos^2 x\, dx = 2\pi, \\
J_3 &= \int_0^\pi \left(\frac{\sin 3x}{\sin x}\right)^2 dx = \int_0^\pi (3 - 4\sin^2 x)^2\, dx \\
&= \int_0^\pi (9 - 24\sin^2 x + 16\sin^4 x)\, dx \\
&= 9\pi - 24\cdot\frac{1}{2}\cdot\pi + 16\cdot\frac{1\cdot 3}{2\cdot 4}\cdot\pi \\
&= (9 - 12 + 6)\pi = 3\pi,
\end{aligned}$$

where we have used the Wallis formula. [See L.-s. Hahn: *Complex Numbers and Geometry,* Mathematical Association of America, Washington, D.C., 1994, p.42.] Thus we conjecture that $J_n = n\pi$.

$$\begin{aligned}
J_{n+1} - J_n &= \int_0^\pi \frac{\sin^2(n+1)x - \sin^2 nx}{\sin^2 x} dx \\
&= \int_0^\pi \frac{[\sin(n+1)x + \sin nx]\cdot[\sin(n+1)x - \sin nx]}{\sin^2 x} dx \\
&= \int_0^\pi \frac{\left(2\sin\frac{(2n+1)x}{2}\cdot\cos\frac{x}{2}\right)\cdot\left(2\cos\frac{(2n+1)x}{2}\cdot\sin\frac{x}{2}\right)}{\sin^2 x} dx \\
&= \int_0^\pi \frac{\left(2\sin\frac{(2n+1)x}{2}\cdot\cos\frac{(2n+1)x}{2}\right)\cdot\left(2\sin\frac{x}{2}\cdot\cos\frac{x}{2}\right)}{\sin^2 x} dx \\
&= \int_0^\pi \frac{\sin(2n+1)x}{\sin x} dx = \pi,
\end{aligned}$$

by Part (a).

APPENDIX E

New Year Puzzles

The author has been sending New Year Puzzles as season's greetings for nearly two decades. As the purpose is to popularize mathematics, these puzzles are not intended to be hard (with some exceptions). Because these puzzles are gaining more and more popularity among the author's friends, we publish them here hoping the readers will do the same.[1]

1985

(a) Note that

$$\begin{array}{rclrcl} 0 & = & (1-9+8)\times 5 & 1 & = & 1-\sqrt{9}+8-5 \\ 2 & = & 1+(-\sqrt{9}+8)/5 & 3 & = & -1-9+8+5 \\ 4 & = & 1\times(-9+8)+5 & 5 & = & 1-9+8+5 \\ 6 & = & 1\times(9-8)+5 & 7 & = & 1+9-8+5 \\ 8 & = & ? & 9 & = & \sqrt{-1+9+8}+5 \\ 10 & = & (1+9-8)\times 5 & & & \end{array}$$

Can you find a similar expression for 8? (Only additions, subtractions, multiplications, divisions, square roots, and parentheses are permitted. The solution is not unique.)

(b) $\square\square\square\square^2 = \square\,19\,\square\square\,85\,\square$.

(c) (i) The square of an integer n starts from 1985:

$$n^2 = 1985\cdots$$

Find the smallest such positive integer n.

(ii) Is there an integer whose square ends with 1985?

1986

Solve the alphametic problem:

```
  HAPPY
-  TIGER
   YEAR
```

under the conditions that

[1] New Year Puzzles, 1985 - 1994, first appeared in Appendix B of author's book: *Complex Numbers and Geometry* (Mathematical Association of America, 1994). They are reproduced here with the kind permission of the Mathematical Association of America.

(a) $TIGER$, being the third in the order of 12 animals (rat, ox, tiger, rabbit, dragon, snake, horse, ram, monkey, cock, dog, boar), the number represented by $TIGER$ divided by 12 gives a remainder 3; i.e.,

$$TIGER \equiv 3 \pmod{12}; \text{ and}$$

(b) as there are 10 possible digits 0, 1, 2, 3, 4, 5, 6, 7, 8, and 9 to fill in the 9 letters that appear in this alphametic problem, there is bound to be one digit missing. However, the missing digit turns out to be the remainder if the number represented by $YEAR$ is divided by 12.

1987

Fill in the blanks with digits other than 1, 9, 8, 7 so that the equality becomes valid:

$$\frac{\square\, 1\, \square\, 9\, \square}{\square\,\square\,\square} = 87$$

1988

(a)

$$\begin{aligned} 1988 &= 12^2 + 20^2 + 38^2 = 8^2 + 30^2 + 32^2 \\ &= 8^2 + 18^2 + 40^2 = 4^2 + 26^2 + 36^2 \\ &= 4^2 + 6^2 + 44^2 = \square^2 + \square^2 + \square^2; \end{aligned}$$

i.e., find another expression of 1988 as a sum of squares of three positive integers.

(b) Show that 1988 cannot be expressed as a sum of squares of two positive integers.

1989

Observe that

$$1989 = (1+2+3+4+5)^2 + (3+4+5+6+7+8+9)^2.$$

Find 4 consecutive natural numbers p, q, r, s, and 6 consecutive natural numbers u, v, w, x, y, z, such that

$$1989 = (p+q+r+s)^2 + (u+v+w+x+y+z)^2.$$

1990

Let

$$P_n = 2191^n - 803^n + 608^n - 11^n + 7^n - 2^n.$$

Then

$$P_1 = 1990, \quad P_2 = 4525260 = 1990 \cdot 2274.$$

Prove that P_n is divisible by 1990 for every natural number n.

1991

(a) In a magic square, the sum of each row, column and diagonal is the same. For example, the figure on the left is a magic square with the magic sum 34. Fill in the blanks in the figure on the right to make it a magic square.

1	8	13	12
14	11	2	7
4	5	16	9
15	10	3	6

	19	91
99		

(b) Can an integer with 2 or more digits, all of which are either 1, 3, 5, 7, or 9 (for example, 1991, 17, 731591375179, 753), be a perfect square?

1992

Choose any five numbers in the figure on the left so that no two of them are in the same row nor the same column, then add these five numbers, you will always get 1992. For example,

$$199 + 92 + 177 + 979 + 545 = 1992.$$

19	92	60	665	470
333	406	374	979	784
94	167	135	740	545
199	272	240	845	650
136	209	177	782	587

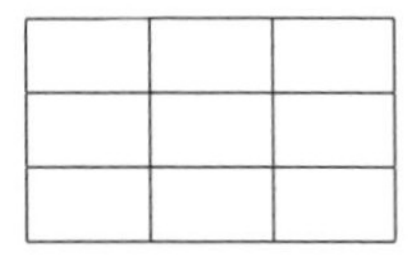

Fill in nine distinct positive integers into the figure on the right such that if you choose any three numbers, no two of them are in the same row, nor the same column, and multiply them together, then you will always get 1992. How many essentially different solutions can you find? (Two solutions are considered to be the same if one can be obtained from other by some or all of the following: (a) rotations, (b) reflections, (c) rearrangement of the order of the rows, (d) rearrangement of the order of the columns.)

1993

Let

$$Q_n = 12^n + 43^n + 1950^n + 1981^n.$$

Then

$$\begin{aligned} Q_1 &= 12 + 43 + 1950 + 1981 = 1993 \cdot 2, \\ Q_2 &= 144 + 1849 + 3802500 + 3924361 \\ &= 7728854 = 1993 \cdot 3878, \\ Q_3 &= 1728 + 79507 + 7414875000 + 7774159141 \\ &= 15189115376 = 1993 \cdot 7621232. \end{aligned}$$

Determine all the positive integers n for which Q_n are divisible by 1993.

1994

We have a sequence of numbers which are reciprocals of the squares of integers 19 through 94:

$$\frac{1}{19^2}, \frac{1}{20^2}, \frac{1}{21^2}, \ldots, \frac{1}{93^2}, \frac{1}{94^2}.$$

Suppose any pair, a and b, of these numbers may be replaced by $a+b-ab$. For example, two numbers $\frac{1}{32^2}$ and $\frac{1}{66^2}$ may be replaced by a single number $\frac{163}{135168}$, because

$$\frac{1}{32^2} + \frac{1}{66^2} - \frac{1}{32^2} \cdot \frac{1}{66^2} = \frac{163}{135168}.$$

Repeat this procedure until only one number is left. Show that the final number is independent of the way and the order the numbers are paired and replaced. What is the final number?

1995

There are exactly 16 divisors of 1995:

1, 3, 5, 7, 15, 19, 21, 35, 57, 95, 105, 133, 285, 399, 665, 1995.

These can be arranged to form a multiplicative magic square; i.e., the product of entries in each row, column, and diagonal is the same.

95	1995	1	21
3	7	285	665
105	5	399	19
133	57	35	15

How many of such essentially different magic squares can be constructed? (Two solutions are considered to be the same if one can be obtained from the other by reflections and/or rotations.)

1996

Observe that 111111111111111111 (an eighteen digit number) is a multiple of 19:

$$111111111111111111 = 19 \times 5847953216374269.$$

(a) Find, if any, the smallest multiple of 96 whose digits are all identical.

(b) What about 1996?

1997

The number 1997 is truly remarkable. Not only is it a prime number itself, but the numbers formed by its first two digits (19), last two digits (97), first three digits (199), and last three digits (997) are all prime numbers.

(a) When did this happen last time?

(b) When will this happen next time?

(c) Are the numbers formed by the middle two digits of the answers to the two questions above prime numbers?

1998

In a multiplicative magic square, the product of entries in each row, column, and diagonal is the same. Fill in the blanks in the figure with divisors of 1998 so that it becomes a multiplicative magic square.

		111	2
			27
18		6	
3	74		

1999

Observe that the set

$$\{1996,\ 1997,\ 1998,\ 1999\}$$

can be split into two disjoint sets, $\{1996,\ 1999\}$ and $\{1997,\ 1998\}$, such that the sum of the elements in one set is equal to that of the other. $(1996 + 1999 = 1997 + 1998.)$

Is it necessary that the positive integer n $(n < 1999)$ be a multiple of 4 for the set

$$\{n,\ n+1,\ \cdots,\ 1998,\ 1999\}$$

to have the same property? (The numbers of elements in two sets need not be the same.)

2000

For two arbitrary numbers x and y, let

$$x * y = x + y + xy.$$

Find three positive integers a, b, c such that

$$a * (b * c) = (a * b) * c = 2000.$$

How many solutions are there?

2001

(a) Does there exist a triple of distinct odd positive integers p, q, r such that

$$\frac{1}{2001} = \frac{1}{p} + \frac{1}{q} + \frac{1}{r}\,?$$

(b) Can p, q, r be distinct even positive integers?

[If possible, try to make the largest denominator as small as possible. Using a computer to solve this problem is like using a sledge-hammer to kill a mosquito.]

2002

(a) In a multiplicative magic square, the product of entries in each row, column, and diagonal is the same. Fill in the blanks in the figure with distinct divisors of 2002 so that it becomes a multiplicative magic square.

	2		
7			11
	13		

(b) Draw a 4×4 grid and place four prime factors, $\{2, 7, 11, 13\}$ of 2002, on any four spaces that are symmetric with respect to, but not on, one of the diagonals such that one row and (therefore) one column are empty.
In this way, we can create 72 different multiplicative magic square puzzles, of which Part (a) is just one example.
Show that each and every one of these 72 new year puzzles for 2002 has a unique solution.

2003

Fill in the blanks with positive integers.

$$\begin{aligned} 2003 &= \square^4 + \square^4 + \square^4 + \square^4 \\ &= \square^2 + \square^3 + \square^4 + \square^5. \end{aligned}$$

Is the answer unique?

2004

(a) Is it possible to express 2004 as the sum of the squares of distinct primes? If so, how many solutions are there?

(b) $\square\square\square\square^2 = 20\square\square\square 04$ (One digit in each blank.)

(c) Does there exist a positive integer whose square ends with 2004; i.e., the last four digits of the number are 2004? If your answer is affirmative, find the two smallest such positive integers.

(d) What if the "square" in Part (c) is replaced by "cube" ?

2005

Fill in the blanks with positive integers.

$$\frac{1}{2005} = \frac{\square^2 + \square^5}{\square^2 + \square^3} - \frac{\square^2}{\square^2 + \square^3}.$$

How many solutions are there?

2006

Place the eight distinct divisors of 2006 at the vertices of a cube in such a way that the product of the numbers at the vertices in any one of the six faces is the same.
How many different solutions are there? (Solutions that can be obtained by rotations and/or reflections are considered to be the same.)

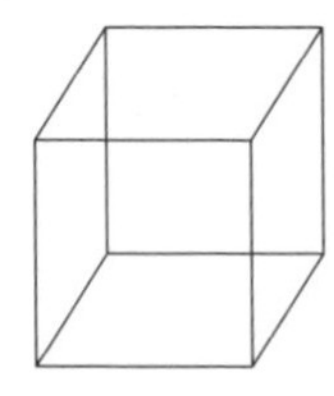

2007

(a) Note that each of the 10 digits $\{0, 1, 2, 3, 4, 5, 6, 7, 8, 9\}$ appears once and only once in numbers $\{10, 2, 953, 6487\}$, and their sum is $10 + 2 + 953 + 6487 = 7452$.
Can you find positive integers such that each of the 10 digits appears once and only once, and the sum of these numbers is 2007?

(b) What if 0 is not allowed?

2008

Let p be a polynomial with integer coefficients. If $p(x) = 2000$ for 3 distinct even numbers, can $p(n) = 2008$ for some integer n? Justify your answer.

2009

(a) Show that there exists a unique polynomial p with integer coefficients satisfying simultaneously the relations

$$\begin{aligned} \sin(2009x) &= p(\sin x), \\ \cos(2009x) &= p(\cos x). \end{aligned}$$

What is its degree? Its leading coefficient?

(b) Find a pair of distinct positive integers p and q such that

$$\frac{p^2 + q^2}{p + q} = 2009 \qquad (p > q > 0).$$

Is the solution unique?

2010

(a) In a multiplicative magic square, the product of entries in each row, column, and diagonal is the same. Given an arbitrary positive multiple of 2010, show that it is always possible to fill in the blanks with positive integers to form a multiplicative magic square whose common product is the preassigned multiple of 2010.

	2	5	
3			
67			

(b) If the common product is 2010^n $(n > 0)$, find the number of solutions in terms of n.

2011

(a) We have a triangle; one of its sides has length $\sqrt{2011}$, and the opposite angle is $\frac{2\pi}{3}$ (i.e., 120^0). If the remaining two sides have integer lengths, find their lengths.

(b) What if the opposite angle is $\frac{\pi}{3}$? Is the answer unique?

2012

Place the six divisors $\{1, 2, 4, 503, 1006, 2012\}$ of 2012 at the intersections of circles such that the product of the integers on any given circle is equal to that of any other circle. How many solutions are there? (Solutions that can be obtained by rotation and/or reflection are considered to be the same.)

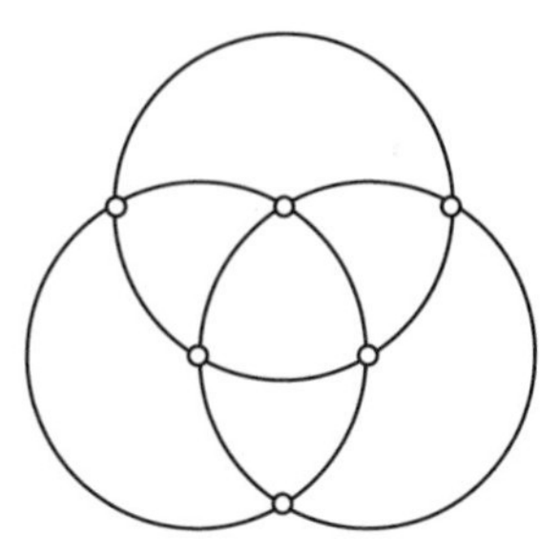

2013

Find all the positive integers n for which the product

$$\left(20^2 + n\right)\left(13^2 + n\right)$$

is a perfect square.

2014

Observe that by placing the integers

$$0, 7, 2, 5, 4, 6, 2, 8$$

around a circle (in this order), we see the positive differences between pairs of adjacent integers form 8 consecutive integers (in some order).

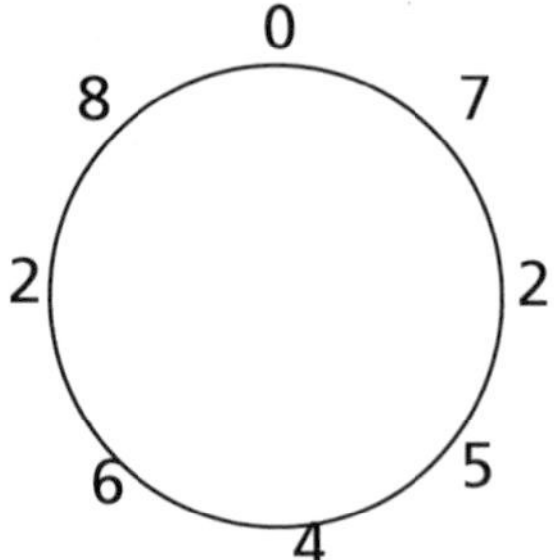

Is it possible to place 2014 integers (repetition allowed) around a circle so that the positive differences between pairs of adjacent integers form 2014 consecutive integers (in some order)?

2015

(a) Find a triple of positive integers a, b, c $(a > b > c > 0)$ such that

$$\frac{abc}{a+b+c} = 2015.$$

(b) Find a triple of positive integers x, y, z $(x > y > z > 0)$ such that

$$\frac{xyz}{yz+zx+xy} = 2015.$$

Is the answer unique in either Part (a) or Part (b)?

2016

Is it possible to arrange the 36 distinct divisors of 2016 in a 6×6 array to create a multiplicative magic square? If so, how many such multiplicative magic squares are there? (Two magic squares are considered to be the same if one can be obtained from other by reflection and/or rotation.)